Statistics
for the teacher

A. C. Crocker

NFER Publishing Company Ltd.

Published by the NFER Publishing Company Ltd.,
Book Division, 2 Jennings Buildings, Thames Avenue,
Windsor, Berks., SL4 1QS
Registered Office, The Mere, Upton Park, Slough, Berks., SL1 2DQ

First published by Penguin Books Ltd., 1969
2nd Impression 1971
3rd Impression 1972
New Edition Published by NFER 1974
© *A. C. Crocker, 1969, 1974*
85633 042 6

Printed in Great Britain by
John Gardner (Printers) Limited, Hawthorne Road, Bootle, Merseyside L20 6JX

Distributed in the USA by Humanities Press Inc.,
450 Park Avenue South, New York, NY 10016, USA

CONTENTS

To
Danny Webster
and
Professor Roy Sommerfeld

PREFACE

STATISTICS IS A SUBJECT with the image of being notoriously difficult both to teach and to understand. Teaching the use of statistics in education to non-mathematicians has posed some particular problems for me, including:

1. How far can I move from strict mathematical language whilst retaining accuracy?
2. Knowing that many people will just want to know 'which knobs to press', how can I put across to them that a knowledge of the processes will help to reduce errors of usage and the acceptance of impossible results?
3. Should I present material in mathematical sequence, or in order of difficulty? What about frequency of use?
4. What should be left out or included?
5. Will a student be able to understand a more advanced text if he has first read my introductory text?

The first edition of *Statistics for the Teacher* was thoroughly tested by many student teachers on different courses between 1969 and 1973, with a view to producing this revised second edition.

I should like to take this opportunity of expressing my gratitude to the following people for their help in producing this book: to Peter Rush, Head of the Mathematics Department, Bede College, Durham, for the many hours given to patient help and advice: to Claude Cunningham, Head of the Mathematics Department, Callendar Park College, Falkirk; to Janet Atkinson for her cheerful acceptance of the typing burden; to Harry Barron, Norman Dodds, Peter Feek, Peter Harris, Christopher Hewitt, David Hughes, Norman Lowdon, Michael Metcalfe, Doug Ridley, Hugh Ridley, Lance Robinson, John Waldren and Christine Davies, my B.Ed. students, for whom this was originally written, for their comments, queries and valuable assistance during the tryout stages. And lastly, my colleague and friend Alan Sanders of Shenstone College for his help, tuition and advice during the last few years.

INTRODUCTION

WHEN I WAS about ten years old I was given an arithmetic work card by my teacher. The last sum read, '$x + 7 = 9$, what is x worth?' I took the card out to the teacher and asked what sort of sum it was. She replied, 'Oh, that's algebra.' Now I had read about the scourge called algebra in the many school stories I loved; I had heard about it from older boys and girls already at the grammar school. Algebra was something terrible. I instantly knew I couldn't do it. At home my brother, aged eight, saw the card and provided me with the answer to the sum. He hadn't heard of algebra.

Many years later the same sort of thing happened when I started to learn some statistics—firstly, I assumed it was beyond me because of what I had heard, and secondly, my assumption was reinforced by some of the most awful teaching that I've ever endured. To my amazement some years later an American professor, Roy Sommerfeld, was able to show me that basically statistical calculations use very elementary algebra and arithmetic. Even today I rarely do any statistical work which requires algebra more complex than that which I had learned by the time I was fourteen.

Provided that you can substitute numerical values in a simple equation involving x or x^2, then statistical calculations are well within your grasp.

Whilst using this book I should like you to remember three important things:

1. It is advisable, on account of the volume of material, to digest the book slowly and by degrees.

2. The end-of-chapter problems will help to consolidate fresh facts.

3. Some statistical terminology is used.

At this point you might well ask, 'Why must there be a new language to learn, why can't we just use simple words that everyone understands?' One of the problems of using words which everyone understands is that we don't all understand the same thing. Take the word 'average'. Try asking people what they mean by average and you'll get answers like the following:

(i) 'Middling, you know!'
mean (ii) 'Half way between top and bottom.'
median (iii) 'My wages are average—same as everyone else gets.'
(iv) 'It's what people say to you when you aren't very good—
mode "Never mind, you're average." '
mode ? (v) 'Take the weather today, it's average. Not good. Not bad.'

When you work out what is meant by these versions you can only come to the conclusion that they don't all mean the same thing. Obviously when using a set of facts we need to be absolutely certain that they mean the same thing, no matter who supplies the data. Hence the need for a precise language.

In education we tend to use statistics:

(i) to predict how likely it is that something will happen and/or
(ii) to check whether a test is any good and/or
(iii) to balance two sets of marks (where for instance one examiner is lenient and another is harsh, in marking the same GCE paper) and/or
(iv) to tell us something about a population from which we have tested a sample.

'So,' you might say, 'that's a use for statistics in education. What about in the ordinary classroom?'

1. I can remember a teacher who regularly claimed that 'the class I've got this year isn't a patch on last year's.' Were his classes progressively worse or was his memory unreliable? A set of scores from a battery of standardized tests given at the start of each year would have shown how true his statement was. Measurement will allow you both to compare *and* to assess more accurately the level of preparation for teaching a subject.

2. Should the form prize go to Johnny because he is brilliant at mathematics (one of the few subjects where it is possible to get 100 per cent if we use traditional methods of testing and marking) and so offsets the weak marks he gets for French? A simple statistical calculation will help the class teacher to balance the value of marks

in a subject allowing high scores with those tending to give lower scores.

3. A knowledge of statistical language and meanings can help the class teacher decide which of a set of standardized tests is best for his purpose, when reading through the various test manuals.

4. To me the most important reason for a teacher to know something about statistics is because he or she is then less likely to accept a statement involving numbers without question, and less likely to be misled by small differences in the scores between two children. Quite recently the following true story was told to me. A pair of twins were given several intelligence tests. Each time twin A scored about two points more than twin B. After five tests it was pointed out that twin A was ten IQ points better than twin B. Suppose the twins had been mile runners and A always beat B by one second. After ten races would we say A is ten seconds faster than B in the mile race? Obviously not. Had the people dealing with the twins known anything about testing or statistics such a case would never have arisen.

5. Many educational publications are containing an increasing amount of statistical evidence which requires a knowledge of statistics if the articles are to be understood.

6. Remember that statistical methods are tools; the skill of the user, in knowing which formula to use and in interpreting the results, is all important.

This book begins with an account of the different forms of average, showing what we may think of as the middle value, and the dispersion of the data around the average; then the use of graphs in displaying numerical information. There are chapters on tests, their reliability and validity and how to analyse the test items; marking and standardized scores; correlation, standard errors and significance. tests. The material is drawn from the classroom and the needs of the teacher who wishes to measure and understand the progress of his or her pupils.

Before going on to the first chapter, I would like to introduce the exact meaning of the word *parameter*.

If we have figures about a whole population we don't call them statistics, we call them *parameters*. (Note: a population is not necessarily large, i.e. if you were interested in a particular class of children and tested all of them, then you have information about that population. You have a set of parameters. If you now use that information to infer something about the whole school population then it becomes a set of statistics.)

Finally, when you have worked your way through this book may I suggest that you obtain a copy of E. F. Lindquist's book *Statistical Analysis in Educational Research*, Houghton Mifflin, 1940. It is readable, interesting and will undoubtedly add much to your knowledge.

SYMBOLS AND ABBREVIATIONS

x	any score
y	any score except x
N	number of people or scores in sample or population
M	mean
$\sum$	the sum of
SD	standard deviation of a sample
Sigma or σ	standard deviation of a population
d	deviation of score from the mean, or difference between two scores
R	reliability coefficient of combined tests, or multiple correlation coefficient
r	zero order correlation coefficient of reliability or validity
SE_{meas}	standard error of measurement
SE_{mean}	standard error of mean
SE_{est}	standard error of estimate (or prediction)
q	quantity
t	(as used in this book) is the number found when calculating the t-test for significant difference between small groups
h	top fraction of class used in item analysis
l	bottom fraction (same size as chosen for h) of class used in item analysis
z-score	a raw score converted to standard deviation value
T-score	standard score based on mean of 50 and sigma of 10
i	interval size when working with grouped data
f	frequency of scores within an interval
σ^2	variance (square of standard deviation for a population)

x^2 chi-square, an index obtained when comparing expected results with actual results

df degrees of freedom

F-ratio number obtained when comparing small groups to see whether at least one of them differs significantly from the total population

A actual result

E expected result

CHAPTER ONE
Central tendency

CENTRAL TENDENCY MEANS the middle score or scores. However, it can be interpreted in three different ways.

1. The Mode   *The score which occurs most frequently*

Do we mean the score which most people get? If we do, we call this the *mode*. For example, eleven people take a test and get the following scores:

$$22 \quad 27 \quad 19 \quad 23 \quad 8 \quad 23 \quad 20 \quad 23 \quad 17 \quad 18 \quad 23$$

What is the mode? In this case it is clearly 23, as four people got that score whilst only one person got any of the other scores. This we call the *modal score*.

2. The Median

Do we mean the score gained by the middle person on a test? If we do, we call this the *median*.

To find the middle person first we have to put the scores in order. The one we usually use is called *rank order*, i.e. we arrange the scores so that they are in order of size.

	Score	*Rank or position*
1	27	Top or 1st
2	23 ⎤	
3	23 ⎥	2nd, 3rd, 4th, 5th or
4	23 ⎥	3·5 average
5	23 ⎦	
6	22	6th
7	20	7th
8	19	8th
9	18	9th
10	17	10th
11	8	11th

Note that where there is more than one person with the same score we take the average (called arithmetical average) of their positions. This is different from the more usual classroom practice, where the teacher would have given the following positions for the same scores:

Score	Position
27	Top or 1st
23	2nd
23	2nd
23	2nd
23	2nd
22	6th
20	7th
19	8th
18	9th
17	10th
8	11th

There are still eleven people and the person who scored 22 marks is still sixth, *but* those who scored 23 are ranked equal second rather than 3·5 as previously shown.

We do not use the second method in statistics, because the effect of equivalent scores all being grouped at the first available rank position would mean that those in the top half would be grouped further from the middle, whilst those in the bottom half would be grouped nearer to the middle. This would make calculations which use rank order (rather than actual scores) subject to serious error.

Now let us get back to the median. The middle person in the above example is number 6. His score is 22. Thus the median score is 22.

When there is an even number of scores, there is obviously no actual middle person. In this case we simply take the score midway between the lowest score in the top half and the highest score in the bottom half.

Score	Rank
9	1
7	2

————————————————————— median

Score	Rank
6	3
4	4

The median rank is midway between second and third place, in this case 2·5. The median score similarly is midway between 7 and 6, in this case 6·5.

Question 1

Put the following scores in rank order and find the median score.

9	5	6	4	3	10	8	2	7	1
19	25	23	27	28	18	20	29	21	30

median score = 24

3. The Mean

The type of central tendency most often used is called the *mean* or arithmetical average. To find this we simply add up all the scores:

$$27 + 23 + 23 + 23 + 23 + 22 + 20 + 19 + 18 + 17 + 8 = 223.$$

and divide by the number of people (11)

$$\text{mean} = \frac{223}{11} = 20 \cdot 27.$$

Usually we use x to stand for a score, N to stand for the number of people, and M to stand for the mean.

This sign $\sum$ means 'the sum of', so the simple formula for calculating the mean is:

$$M = \frac{\sum x}{N}$$

Follow this example: calculate (i) the median and (ii) the mean annual wage for the following five men.

A	£7900
B	£2000
C	£1100
D	£1000
E	£500

(i) As they are already in rank order (order of size), the median wage is that of C = £1100 per annum.

(ii) Using the formula for the mean

$$M = \frac{\sum x}{N};$$

$\sum x = 7900 + 2000 + 1100 + 1000 + 500 = 12{,}500,$
$N = 5,$
$$M = \frac{12{,}500}{N} = £2500 \text{ per annum.}$$

As you can see from this example, the mean wage is more than that earned by four of the men because of the high wage earned by one man. We call this result *skewed*. Because of the effect of a few high salaries when national average earnings are quoted, the median earnings rather than mean earnings are usually given.

Question 2
Find the mode and mean of the following numbers:

2 7 3 4 9 8 9 6 9 4 1 9 5

Question 3
Put the following numbers in rank order and find the median score:

19 27 19 8 31 9 22 11 14

Question 4
When we know figures about a whole population, we don't call them —, instead we call them — or facts.

Question 5
Where the mean is numerically a long way from the median for a set of figures, it is likely that a few figures are causing the mean to be —.

Question 6
Take a look at the various descriptions of 'average' on page twelve. Are they referring to mode, mean or median in each case?

Usually when we have a class of children, we are not so fortunate as to have only 12 or so scores to work out. Try this last problem based on a full-sized class.

Question 7

The following are class 2B's scores for English for the summer term 1968.

Betty	13 207	Maureen	204 14	Jim	32 186	Dick	191 27
Jean	23 196	Paula	3 211	Fred	21 198	Roger	208 10
Freda	17 201	Anne	1 287	Charlie	15 203	Stan	9 210
Gerty	29 189	Jill	18 200	Bert	2 281	Reggie	195 24
Nora	26 193	Linda	13 205	Mike	5 216	Paul	200 19
Norma	3 274	Barbara	195 25	Tom	7 214	George	209 9
Joan	6 215	Cathy	71 187	Sid	22 197	Ted	4 219
Ivy	13 182	Delia	30 188	Norman	189 28	Nigel	208 10 11
Iris	199 20			Peter	202 16		

(a) Put them in rank order.
(b) Calculate the mean score.
(c) Calculate the median score.
(d) Calculate the modal score.
(e) Does the mean or median score give a better picture of the average performance in the class?
(f) Why have you made the decision in (e)?

$\Sigma x = 7057$

$\therefore m = 207.6$

1	287	19. 200	33. 186
2	281	19.5	34. 182
3	274	20 200	
4	219	21. 199	$\Sigma x = 559$
5	216	22 198	34.368
6	215	23 197	
7	214	24. 196	mean = 208.2
8	211	25 195	
9	210	25.5	368
10	209	26 195	2717
11	208	27. 193	3974
12	208	28 191	6059
13	207	29 189	
14	205	29.5 189	
15	204	30. 189	
16	203	31 188	
17	202	32. 187	
18	201		

3974   2717

1856   1677   1888   1638

CHAPTER TWO
Dispersion

DISPERSION IS THE spread or scatter of scores. The mean by itself tells us very little about a set of scores. Take, for instance, the following scores all with the same mean of 6.

$$
\begin{array}{llllll}
\text{(a) 6} & 5 & 6 & 7 & 6 & \text{mean} = 6 \\
\text{(b) 1} & 6 & 11 & 2 & 10 & \text{mean} = 6 \\
\text{(c) 1} & 2 & 3 & 4 & 20 & \text{mean} = 6 \\
\end{array}
$$

In group (a) the scores are very close to the mean, in group (b) they are much more spread out, whilst in (c) the mean is raised considerably because of one score. Clearly we need to know a bit more about a set of scores than the mean can tell us. There are several ways of doing this.

1. The Range
The simplest way to find out more about a set of scores is to record the *range* from the bottom score to the top score. In example (a) above, the range is from 5 to 7 *inclusive* of both scores, i.e. the range is 3. In example (b), the range is from 1 to 11, i.e. 11, and in example (c), the range is from 1 to 20, i.e. 20.

If we reported that a set of five scores had a mean of 6 and a range of 3, it would be very easy to see that the scores were less spread out than a set of five scores with a mean of 6 and a range of 20. However, this still doesn't tell us *how* the scores are spread out.

2. The Standard Deviation
Generally speaking, the way we show how much scores are spread out is by quoting the *standard deviation*. This is a measurement which shows how much the scores, as a group, deviate from the mean. In other words, it shows how much *all* of the scores are spread out and not just the gap between the highest and lowest scores (which is all that the range tells us). The words 'standard deviation' are usually

shortened to either SD when we are talking about a sample of scores (i.e. a statistic), or σ (sigma) when we are talking about a population (i.e. a parameter).

One of the simplest formulae for calculating standard deviation is:

$$\sigma = \sqrt{\left(\frac{\sum d^2}{N}\right)}$$

where $\sum$ is 'the sum of', d is the deviation of each score from the mean, and N is the number of scores.

Let us take the example (b) again. Here the scores were 1, 6, 11, 2, 10 and the mean was 6.

(i) Put the scores in order down the page.

1
6
11
2
10

(ii) Now calculate how much each deviates from the mean.

1 deviates from 6 by 5
6 deviates from 6 by 0
11 deviates from 6 by 5
2 deviates from 6 by 4
10 deviates from 6 by 4

(iii) Now square each of these deviations.

$5^2, 0^2 = 0, 5^2 = 25, 4^2 = 16, 4^2 = 16$

(iv) Add them all up.

$25 + 0 + 25 + 16 + 16 = 82.$

This is the sum of the squared deviations,

i.e. $\sum d^2 = 82.$

(v) Now divide by the number of scores.

$$N = 5,$$

$$\frac{\sum d^2}{N} = \frac{82}{5} = 16\cdot4.$$

(vi) Lastly, find the square root of 16·4.

$$\sqrt{16\cdot4} = 4\cdot05.$$

The standard deviation of the scores in example (b) is 4·05.

In actual practice we would make the scores up into a table. This way it is easier for us to see if we've left a score out and for someone else to see exactly what we have done.

The table for the above example would look like this:

Score	Deviation from mean	Deviation squared
x	d	d^2
1	5	25
6	0	0
11	5	25
2	4	16
10	4	16
		$\sum d^2 = 82$

In many text books deviation from the mean will be given as $x - M$ or $x - \bar{x}$. This means that, when scores are below the mean, the deviation will come out negative. However, squaring removes the minus sign.

Question 8

Calculate the standard deviation in example (a) of this chapter.

Question 9

Calculate the standard deviation in example (c) of this chapter.

The formula which I have given you produces quite accurate results for a set of more than thirty scores. However, it does have to be modified in one case: when we are estimating the standard deviation of a population from a sample.

The formula becomes:

$$\text{Standard deviation} = \sqrt{\left(\frac{\sum d^2}{N-1}\right)}$$

This means that the standard deviation will be bigger than if we used N as the denominator. In the case of small samples, the modified formula gives us a closer approximation to the sigma for the population we are trying to estimate than does the original formula.

Question 10
Now recalculate the standard deviation for example (c) assuming it to be a sample from a larger population.

In a normal distribution roughly 68 per cent of a population will get scores (weight or height or whatever we have measured) within one standard deviation of the mean. For instance, the Stanford–Binet Intelligence Test (1960 revision) has a standard deviation of 16. So we know that approximately 68 per cent of the population will get an IQ score within 16 points of the mean (which is 100),* i.e. between 116 above and 84 below the mean. Approximately thirty-two per cent of the population will have IQ scores *more than* one standard deviation away from the mean. In fact 16 per cent will have IQ scores above 116 and 16 per cent will have scores below 84. Approximately ninety-five per cent of the population will get scores within approximately 1–96 standard deviations from the mean, that is, an IQ between 68 and 132. This leaves only 2½ per cent who score

---

* Otis and Stanford–Binet in common with most modern tests of IQ provide deviation IQs, i.e. they are standardized scores around the chosen mean of 100 (just as the McCall *T*-scores fluctuate around the chosen mean of 50; see Chapter 9).
  Older tests of intelligence were based on the ratio of mental age to chronological age. This fraction was then multiplied by 100. In theory, an average child would have the same mental age as his chronological age, so the fraction $\frac{\text{M age}}{\text{C age}}$ would be 1.
  $1 \times 100 = 100 =$ the average IQ.

so low that they are below an IQ of 68 and $2\frac{1}{2}$ per cent so high that they score above 132.

How does this knowledge help us?

(i) We can compare an individual score with the population and see whether it is above or below the mean. And we can also see *how far* the score deviates in terms of the population.

(ii) Where different tests have different standard deviations, we can still compare a person's position within the population. Take for instance two IQ scores. George scored 130 on the Otis Gamma IQ test and Bill scored 130 on the Stanford–Binet IQ test. Who has the higher IQ? The standard deviation of the Otis is 10. As mentioned before the standard deviation for the Stanford–Binet is 16. Therefore George has scored three standard deviations above the mean. He is in the top 0·10 per cent of the population for IQ. Bill is not quite two standard deviations above the mean. He is in the top $2\frac{1}{2}$ to 3 per cent of the population but on this comparison would appear to have a lower IQ than George.

(iii) Another piece of information which we can get from the standard deviation is a mental picture of how the scores are spread out around the mean. If the standard deviation is small *relative to the range* (i.e. the standard deviation is clearly less than one sixth of the range), then we know that the scores tend to be heavily clustered close to the mean, with only a few scores a long way from the mean.

If on the other hand the standard deviation is large *relative to the range* (i.e. the standard deviation is clearly larger than one sixth of the range), then we know that the scores will tend to be widely spread over the whole range.

In actual practice this sort of picture can often be seen from the raw scores without going to the trouble of calculating the standard deviation. The value of the standard deviation lies more in its integral part of many other essential statistical calculations.

We talk of deviations above the mean as being positive or plus, and deviations below the mean as being negative or minus.

Question 11

Given a mean of 50 and a standard deviation of 10, how many standard deviations away from the mean are the following scores:

(a) 60   (b) 70   (c) 45   (d) 31   (e) 72?

Standard deviations can be quoted instead of raw scores; thus for 11(d), we could report the score as $-1\cdot9$ standard deviations. Scores reported in this way are called z-scores. We can tell immediately we see a z-score whether a person is above or below average and also *how far*. Ninety-nine point seven per cent of the population fall within $\pm\,3$ standard deviations of the mean (3 people in every 1000 will score outside the $\pm\,3\sigma$ limits).

If it is required to convert a z-score to a percentile rank, this is easily done by referring to Table 10, page 136. Table 10 is based on the normal curve. Percentages and the normal curve are dealt with in Chapter 3.

For example, suppose Johnny goes home and tells his mother that he got 70 per cent today for history. Would she be pleased if she knew that 70 per cent was the bottom mark and that everyone else got above 85 per cent? Clearly a mark by itself tells us little.

However, it is also true to say that Johnny's mother would be pretty puzzled if he came home and said, 'Teacher was pleased with me today. I got a score of plus 2.'

There are other ways of measuring the spread of scores: two are the inter-quartile range and mean deviation. However, in practice these are not often reported, and so I shall not cover them in this book. One which is often used, however, is the measure of *variance*. For our purposes it is necessary to know a little about variance so that we may find out whether the difference between small samples is significant. This is dealt with in Chapter 11. The variance for a single sample or for a single population is exactly the same as the standard deviation squared, and so for simplicity we use the same symbols.

Variance $=\sigma^2$ or SD^2

Variance is dealt with in Chapter 11.

Question 12
Given a mean of 30 and a standard deviation of 5, convert the following raw scores to z-scores:

(a) 35   (b) 40   (c) 25   (d) 20   (e) 16   (f) 30.

Question 13
What percentage of the population have z-scores at or below:

(a) $+3$ (b) $+2$ (c) 0   (d) -1?

Question 14

The following are scores for class 4A in history and geography.

(a) Calculate the mean and standard deviation for both sets of scores.

(b) Decide in which subject Percy has made the better performance, giving your reason.

Name	Geography mark	History mark	Name	Geography mark	History mark
100. Percy	70 +10	70	121 Colin	49 −11	63
4. George	62 +2	75	4 Mike	58 −2	68
900 Harry	30 −32	62	36 Doug	54 −6	69
169 Bill	47 −12	72	64 Phil	68 +8	69
961 Tom	91 +31	71	1369 Nick	97 +37	70
484 Bert	82 +22	68	16 Mark	64 +4	66
100 Norman	70 +10	69	441 Matthew	81 +21	73
16 Fred	64 +4	74	100 Luke	50 −10	67
1 John	59 −1	66	324 Basil	42 −18	68
169 David	73 +13	65	64 Reg	68 +8	65

2904   648 |+48| 692   2539   631 +38   678

            631   692

Geog Total $\dfrac{1279}{20.}$ Hist. total $\dfrac{1370}{20.}$

$M = 63.95$ $M = 68.5$

let $M' = 60$ Calculator

$\sum d^2 = \begin{array}{c} 2904 \\ 2539 \\ \hline 5443 \end{array}$ History : $\sigma_n = 3.88$

                                    $\bar{x} = 68.5$

$\dfrac{\sum d^2}{N} = 272.15$

$\left(\dfrac{\sum d}{N}\right)^2 = \left(\dfrac{+79}{20}\right)^2$ Percy = 70 in both

               $= 3.95^2$ Geog. = $\sigma_n = 16.017$

               $= 15.60.$ $\bar{x} = 64.$

$\therefore S = 16.017.$ Does better in history ∵

                    he is $\dfrac{1.5}{3.88}$ above mean $= .44$

                in Geog he is $6/16$ above $\bar{x} = .375$

CHAPTER THREE
Graphs and curves

IT IS POSSIBLE to plot information in a variety of ways depending on what information we require. Usually when we plot a graph we tend to group information together. For example, if we plot a graph of height against weight for men over fifty years old, we are likely to have a group for men 5 feet 6 inches tall, another for men 5 feet 7 inches tall, etc. However, we need to ask, 'Does 5 feet 6 inches mean anyone between 5 feet 5·5 inches and 5 feet 6·4 inches, or does it mean those *exactly* 5 feet 6 inches tall?' Usually in graph work we tend to take the first meaning. It is important to remember that the bigger we make the category the more information may be hidden within that category. For example, the categories 'men over 4, 5, 6 and 7 feet' would put most men in the 5-foot group but would not show at all how they were spread within the group.

Where information is grouped together in this way, it is known appropriately as *grouped data*. Where each piece of information is taken separately at its own value, this is called *continuous data*. We tend to assume the data is continuous when we calculate from a formula rather than from a graph or grouped data.

One of the simplest forms of graph is the *block diagram* or *histogram*.

Suppose you stood on the corner of your local high street and measured the height, to the nearest tenth of an inch, of the first 1000 men that came past and you got the following results:

Height	Number measured
5′ 4″ and below	50
5′ 4·1″ to 5′ 6″	100
5′ 6·1″ to 5′ 8″	200
5′ 8·1″ to 5′ 10″	300
5′ 10·1″ to 6′ 0″	200
6′ 0·1″ to 6′ 2″	100
6′ 2·1″ and above	50

Our graph is drawn by plotting the number of men in each group against the groups as follows:

Histogram of heights of 1000 men (fictitious example)

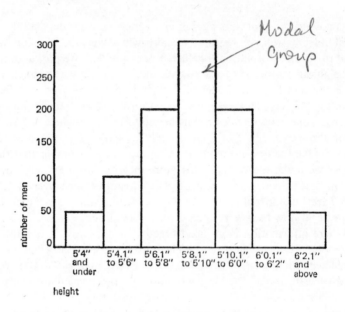

Thus we have a block representing each group. The height of each block shows the number of people in each group. Incidentally the tallest block is called the *modal group*.

In the above example the modal height would be 'between 5 feet 8·1 inches and 5 feet 10 inches'. In simpler language, more people will be between 5 feet 8·1 inches and 5 feet 10 inches than will be in any other height group.

Suppose that we would like to have a smooth curve instead of the step-like histogram. We get this by assuming that all scores within a group fall at the midpoint of that group. For example, take the group of 100 men whose heights were between 5 feet 4·1 inches and 5 feet 6 inches. We assume that they are all 5 feet 5 inches tall. Our histogram transforms from this:

to this:

to this:

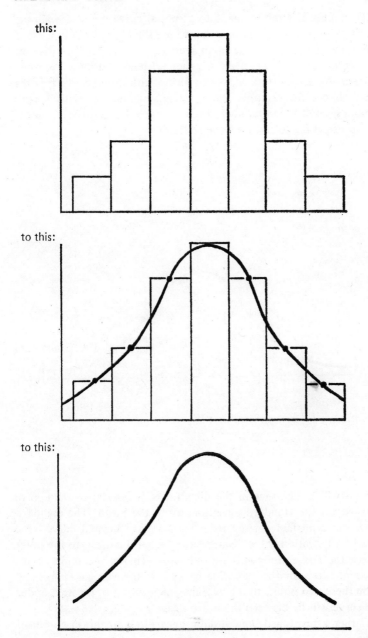

This curve is close in shape to a special curve called the *normal curve*. It so happens in nature that where things vary about a mean (i.e. some things are smaller than average, some larger than average and others close to the average), they all tend to follow the same pattern. As a result we usually finish up with a curve which is the same shape; for example, when we are measuring men's height, annual rainfall over a number of years, or the time it takes a random group of people to run up some stairs.

The normal curve

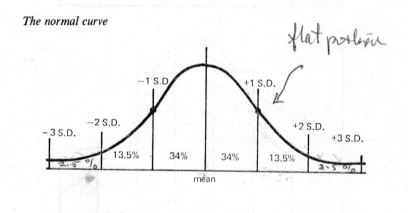

You will notice that in the diagram of a normal curve I have marked off the standard deviations from the mean. The standard deviation is in fact derived from the standard normal curve. One standard deviation is the distance away from the mean to the point where the curve stops curving outwards. This is the point of inflection. I have marked it with a heavy dot. If a vertical is dropped from these two points to the base line, 68 per cent of the area under the curve will be between these two lines.

Earlier I mentioned skewed data. Let us look at this in graphical form.

1. Skewed curve

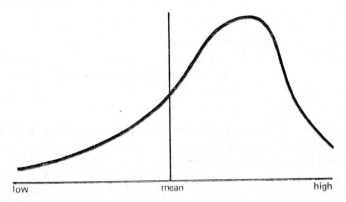

1. This is where a few extremely low scores lower the mean and on the graph they drag the left-hand side out into a tail. Appropriately it is called a *left-handed tail* or left-handed skew.

2. Skewed curve

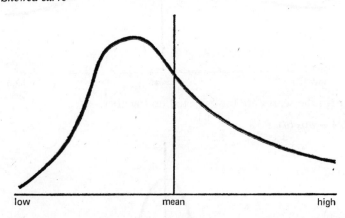

2. This is where a few extremely high scores raise the mean and on the graph they drag out the right-hand side into a tail. This is called a *right-handed tail* or right-handed skew. Examples of skewed distributions can be seen in selection for various jobs or professions, or in Education.

1. The weights of flat-race jockeys must be low and are mostly between six and eight stone. A very few will be below six stone,

thereby dragging the mean down. Graphically this will make a left-handed skew.

2. An example of a right-handed skew is the height of men in the police force. Men have to be over 5 feet 8 inches tall to get in. Most will be between 5 feet 8 inches and 6 feet; fewer will be between 6 feet and 6 feet 6 inches. Thus the tail can only extend one way thereby raising the average.

Kurtosis

This is a special form of distortion of the normal curve. Basically the mean and median coincide. But either (i) the scores are too spread out across the total range;

3. Kurtosis (i)

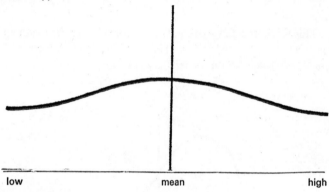

or (ii) the scores are too bunched up together.

4. Kurtosis (ii)

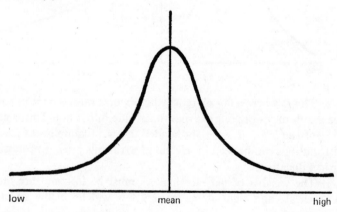

Kurtosis then, differs from skew in that scores are distributed symetrically about the mean which, similarly to the normal distribution curve, is exactly the same as the median.

An example of kurtosis is the height of ballet dancers. The limits set by ballet companies are very narrow—something like 5 feet 3 inches to 5 feet 6 inches. Only rarely will a ballet dancer be outside these limits. This produces a very marked peak and virtually no tails at all.

Let us now look at a third method of plotting information on a graph. This method is often used with examination results where the cut-off point between passing and failing is to be illustrated. It is called the *ogive* or S-curve or cumulative frequency curve. In this type of graph we plot *all* of the scores below a certain point against that point. For example, take the data on page 30 from which we plotted our histogram. In the case of the ogive we would say:

1. How many men are less than 5 feet 4 inches tall? Answer: 50.

2. How many men are less than 5 feet 6 inches tall? Answer: 100 + 50 = 150, i.e. we count *all* men shorter than 5 feet 6 inches tall, not just those in the 5 feet 4·1 inches to 5 feet 6 inches bracket.

Let us plot an ogive using the data on page 30. Firstly we need to rewrite our numerical information as follows:

Original data

Height	Number measured
5′ 4″ and below	50
5′ 4·1″ to 5′ 6″	100
5′ 6·1″ to 5′ 8″	200
5′ 8·1″ to 5′ 10″	300
5′ 10·1″ to 6′ 0″	200
6′ 0·1″ to 6′ 2″	100
6′ 2·1″ and above	50

Data rewritten

Height	Number measured
5′ 4″ and below	50
5′ 6″ and below	150
5′ 8″ and below	350
5′ 10″ and below	650
6′ 0″ and below	850
6′ 2″ and below	950
Group measured (including those above 6′ 2″)	1000

Now we can plot height against number of people below that
height as follows:

The ogive

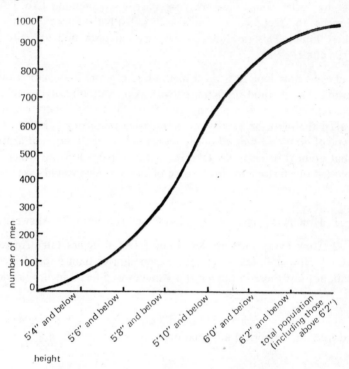

This way of plotting information allows us to divide the population into equal parts without worrying about the scores people
have got. Thus if we wish to separate the population into four equal
groups or quarters, we simply divide the number of people by 4.
In the above example $N = 1000$. Therefore, each quarter of the
population would contain $1000/4 = 250$ people. If we draw a line
on the graph above the 250th person we call this line the *first quartile*.

We can do the same thing in tenths of the population. Here the
cut-off lines are known as *deciles*. Ten per cent of any population is
below the first decile. Ninety per cent of the population is below the
ninth decile.

Below, the graph of heights is shown again, this time divided into deciles and quartiles

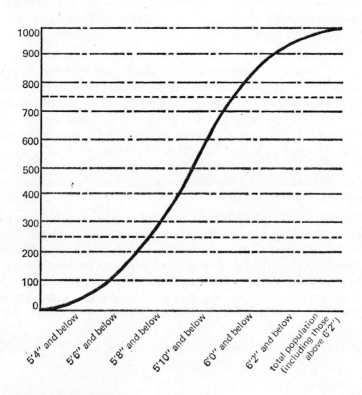

line above 1000	tenth decile, fourth quartile
line above 900	ninth decile
line above 800	eighth decile
line above 750	third quartile
line above 700	seventh decile
line above 600	sixth decile
line above 500	fifth decile, second quartile and median
line above 400	fourth decile
line above 300	third decile
line above 250	first quartile
line above 200	second decile
line above 100	first decile

The usual method of dividing a population into parts is to draw a a line across the graph *above* each 1 per cent. These are called *percentiles*. It is possible to be below or above the first percentile, below or above the ninety-ninth percentile but obviously *never* above the hundredth percentile.

Percentiles are most useful when we are trying to see the value of someone's score on a particular test. For example, what does John's IQ score of 116 mean? If we ask what percentile rank does that place John above and are told 'above the eighty-fourth percentile', we can then see John in relation to the population upon which the test was standardized. If we are told, 'Your new baby girl is nineteen inches long', we may ask, 'Is that big or little?' If instead we are told, 'Your baby girl is below the sixteenth percentile for length', then we know she is small compared with other babies. We can always measure her length for ourselves anyway.

If we want to find what percentile ranking a score falls above or below, we simply draw a line from that score to the curve and then draw the horizontal across to see where it lands on the vertical scale. In examinations, the pass mark is determined by the proportion of the population which the examiners decide to pass, e.g. if they decide to allow 69 per cent to pass, then they draw a line across at the thirty-first percentile, until it reaches the curve, drop the perpendicular on to the marks line and read off the pass mark.

Try the following examples.

Question 15

Given the ungrouped data on page 35:

(a) Sort it into suitable groups.

(b) Construct an ogive.

(c) Draw on the graph the second decile, third quartile and ninety-fifth percentile.

(d) What would the pass mark and credit mark be if 66 per cent were allowed to pass the examination and 12 per cent were given credits?

Data

Marks for history examination, fourth form, St Christopher's School, summer 1968.

John	19	Willie	51	Sheila	38	Penny	60
Fred	27	Joe	80	Jane	48	Sue	70
Peter	4	Len	35	Betty	46	Rita	40
Bill	82	George	41	Freda	88	Laura	44
Reg	59	Mike	55	Norma	32	Kate	55
Norman	63	Harry	65	Maggie	53	Celia	65
Tony	47	Cedric	28	Meg	56	Jean	70
Arthur	73	Cecil	59	Martha	58	Joan	91
Claude	38	Henry	49	Brenda	61	Daisy	68
Basil	22	Dick	38	Ann	63	Esther	33
Humphrey	67	Philip	76	Janet	57	Gert	41
Ray	61	Mark	63	Jill	16	Hilda	49
Bert	59			Dot	73		

One thing worth noting about percentiles is the fact that although the number of people between each percentile is the same, the score ranges vary. This is because more people get scores near to the mean score than near the extremes. For example, if the range of marks is 0 to 100 with a mean of 50, we would expect more people to score between 45 and 50 or 55 and 60 than between 5 and 10 or 85 and 90. In other words the same number of people will cover a wider range of scores at the extremes than near the mean.

CHAPTER FOUR

Reliability and validity

Reliability
WHEN WE GIVE a test to a person or group of persons we need to be concerned with many things.

1. How *consistent* is the test? If we gave it again to the same group would we get the same sort of results?
2. If a different person marked it would it automatically give the same marks?

These questions are important whether the test is a standardized, published test or a home-made, teacher test. The measure of consistency that a test has is called *reliability*. The degree to which it is impossible for different people to give different marks for the same piece of work is a measure of the test's *objectivity*. A truly objective test is one which can only be marked in one way. As a result these tend to be the short-answer type of test where a question is asked (or a statement made) and the correct answer is chosen from a group of answers, or where a testee decides whether a statement is true or false.

A further treatment of objective test questions will be given in Chapter 8. Here it is sufficient to point out that it is usually easier to obtain the same results time after time with an objective test than with an essay test. The amount of ability to reproduce accurately the same results again and again is reflected in the reliability figure.

Reliability is tested by several methods.

The test retest
Here the same test is given twice with a gap of time intervening and the scores on each test compared. If it is children who take the test, it may prove necessary to adjust the scores for improvement caused by the extra month of schooling. A simple method is to compute the two averages and remove the difference from everyone's second score. Another way with large groups is to sort the children into monthly sub-groups, find the average score for each sub-group and remove

the one-month average difference from each score. For example.

(a) Average score, boys aged 9 years 2 months = 217
(b) Average score, boys aged 9 years 3 months = 219
(c) Average score, boys aged 9 years 4 months = 222

In this case we would remove 2 points from each boy who was 9 years 3 months on the second testing, and 5 points from each boy who was 9 years 4 months on the second testing. Then we could compare the two sets of scores.

If we have just tested our own class then the first method would be the better one to adopt. The second method would probably mean (for a class of thirty or forty children) that only three or four children had their birthday in each month. To assume that any increases in scores were due to chronological advantage (i.e. luck of being older) would be very dangerous with such a small sample. The second method is a better way of computing *age allowances* when a large group have taken a common test, e.g. after a whole town's eleven-year-olds have taken the eleven-plus exam.

Parallel forms of a test

In this case two similar versions of a test are given and the results compared. The closeness of the relationship is again taken as being a measure of the reliability of the test. Here there is no need to balance for the effect due to increase of age as both tests can be given close together.

Split-half method

In this method the test is split into two halves, each question is balanced against another of equal difficulty. If the test is of gradually increasing difficulty then it is too simple to take the odd and even questions. Instead we arrange the two halves as follows:

question	1	4	6	7
against	2	3	5	8

The sequence is repeated every eight questions. This way number 2, slightly more difficult than number 1, is balanced with 3 which is easier than number 4, and so on.

Within reason the longer the test the more reliable it is. This is because a longer test gives us more information than a short test, and also small differences between testees are more likely to show if there are many questions rather than a few.

A rough way of calculating the reliability of a whole test from the split-half reliability is with the following formula:

$$R = \frac{2r}{1 + r}$$

where R = reliability for the whole test
and r = relationship between the two subtests.

At this point you are probably asking 'How do I find out the relationship between the two tests, whether they are test–retest, parallel form or split-half?' In statistics we tend to use the word *correlation* rather than relationship. Some methods of calculation are outlined in Chapter 6.

The correlation coefficient of reliability has to be high for a test to be of use to us, or for a marking system to be above suspicion. Any test with a correlation coefficient of reliability between + 0·9 and — 0·9 is suspect. Take the following example from an article in *Educational Research*, November 1964.

Two markers had each marked 100 people in 'spoken English'. Both were regarded as having failed the bottom 25 per cent. The reliability correlation between the two markers was + 0·75.

		Marker Z		
		Top 75%	Lower 25%	Total
Marker Y	Top 75%	66	9	75
	Lower 25%	9	16	25
	Total	75	25	100

The reliability coefficient of a test is an estimate of how much test results may vary if the testing is repeated. The table shows this estimate clearly. It suggests that there will be a misclassification to the extent that (there being 100 candidates) of marker Y's pass list of 75, marker Z would fail 9; and of marker Z's pass list of 75, Y would fail 9. Thus from an entry list for a test in spoken English of 100 candidates, $9 + 9 = 18$ will be misclassified. This is a rather unsatisfactory state of affairs and suggests that a group of two, three or four assessors is likely to provide more reliable assessments than a single assessor.

Many tests however have *validity coefficients* below + 0·9 or above — 0·9 but are still of great use to us. Let us now take a look at validity and see what that is all about.

Validity

This means truth or fidelity. In short—does something measure what it claims to measure?

When we give a test we must always ask, 'What are we *really* trying to measure?' For example, suppose two boys do an arithmetic test. Johnny gets all ten sums right. Freddy gets eight sums right. Johnny is untidy so the teacher takes 3 marks off his total. The scores now read:

> Freddy   8 out of 10
> Johnny   7 out of 10
> Therefore Freddy is top for arithmetic

Is this valid? Is Freddy better than Johnny at arithmetic? Clearly the answer is no. The teacher has muddled presentation with process, but has *apparently* only given marks for process. This example is not included in order that presentation should be decried *but* in the hope that it may help you to see that it is necessary to look at it as a separate issue when it comes to awarding marks.

Let us now look at the various forms that validity can take in statistics.

1. Predictive validity

We can't measure future performance; we can only measure present performance and predict future results.

The eleven plus is a predictive test which, despite all the controversy about it, is nevertheless the best single predictor (obtainable at the age of eleven) of academic performance at sixteen plus and eighteen plus. Whether there should be selection at eleven is a different thing altogether. Indeed it is extremely sad to see that many local education authorities have kept selection at eleven plus but abandoned the eleven-plus exam. The predictive validity of primary school headmasters' assessments is known to be lower than that of the test, and more likely to be swayed by subjective red herrings.

If we use a test to measure someone's current level of performance, then that test is called an *achievement test*. If we use the results to predict future performance then we call it an *aptitude test*.

2. Concurrent validity

Here we are attempting to estimate something which we cannot measure directly. For example, we can't measure how hot it is directly, but we can measure how much a column of mercury has expanded and then *infer* the temperature from that.

3. Construct validity

A construct is an hypothetical attribute such as beauty or intelligence. You can't see 'intelligence', you can only see the results of intelligent action. Similarly, with beauty what we see is something which we believe is beautiful. Often a construct is argued over because there is no concrete frame of reference. How many of us have disagreed with the judges' decision in a beauty contest?

4. Content validity

Is each question in a test actually testing what the whole test claims to test? Many questions in a physics test are really testing ability to juggle with symbols and numbers, rather than testing a knowledge of physics. For example:

Given the following formula and information, calculate v.

$$v = u + at,$$
$$v = ?$$
$$u = 0,$$
$$t = 3,$$
$$a = 32.$$

$$v = 0 + 3 \times 32$$
$$= 96.$$

This is often dressed up something as follows: If a stone is dropped from rest over the edge of a cliff, calculate its velocity after three seconds.

Acceleration due to gravity $= 32$ ft/s²,
$$v = u + at.$$

Validity is checked by finding the relationship (correlation) between what we have measured and one or more of the following:

(a) actual future performance
(b) expert opinion
(c) results of another test of known and accepted validity.

Validity correlations also fall in the range — 1·0 through 0·0 to + 1·0. Unlike reliability coefficients they can be much lower and still tell us a lot that we would like to know. The following table is a rough guide.

R from   0·00 to ± 0·20—indifferent and negligible
R from ± 0·20 to ± 0·40—low, present but slight
R from ± 0·40 to ± 0·70—substantial and marked
R from ± 0·70 to ± 1·00—high, rising to perfect

Finally it is worth pointing out the following:

1. A test *can* be reliable but invalid. (Hundred yards races are a reliable way of showing who is fastest in a class but it would not be valid to use the results as a predictive measure of age of onset of baldness, for example.)

2. If a test is not reliable then results obtained with it cannot be valid. For example, a test can be reliable without being valid but it cannot be valid without being reliable. A good illustration of this is given by J. R. Amos, G. Brown and O. C. Mink (see p.144), who point out that a rifle carefully held might hit the same point consistently. This is reliability. If the bullet also hits the target you are aiming at, it is valid. A rifle needs to be both reliable and valid in its application to be of any value.

CHAPTER FIVE
Standard errors

As MENTIONED IN Chapter 2, when predicting the population σ from a sample standard deviation we have to be cautious. This is equally true when reporting a score or mean, or predicting a future score.

Reporting Scores and Means
Standard error of measurement

Suppose a girl takes an IQ test and scores 119. If she took the test again or took a similar test, or had taken it feeling unwell the first time, would we still expect the same score? Would she still make exactly the same chance mistakes such as putting a tick in the wrong box whilst knowing the right answer? Of course we know that the chances are that she wouldn't get the same score on another occasion. However, she would be *likely* to score fairly close to her original score; how close depends on chance for the population as a whole. Because it depends on chance we find that the *standard errors* fluctuate according to a definite pattern. The pattern already described is that of the normal distribution or normal curve, i.e. most scores will only vary a small amount and only a few will vary widely. This variation is called the *standard error of measurement* or SE_{meas} for short. When the range of possible marks, and thus the standard deviation, is large, then the SE_{meas} will also be relatively large.

Only one SE_{meas} is calculated for any one set of scores and then applied to any individual score within that group. Because errors tend to produce the pattern of the normal curve, in the same way that scores do, we are able to make confident statements about a person's score. For example:

(a) 68 per cent of the population will have a true score within 1 SE_{meas} of their actual score.

(b) 95 per cent of the population will have a true score within approximately 2 SE_{meas} of their actual score.

(c) 99·7 per cent of the population will have a true score within approximately 3 SE_{meas} of their actual score.

Putting this another way, we can be confident that in 68 per cent of all cases a person's true score will be within 1 SE_{meas} of his actual score, whilst only five times out of 100 will the discrepancy be more than 2 SE_{meas}.

The SE_{meas} is calculated very simply once the standard deviation of the group taking the test is known.

$$SE_{meas} = \sigma \sqrt{(1 - r)},$$

where σ = standard deviation for group
and r = reliability figure for test (we deal with the calculation of reliability in Chapter 6).

For example, suppose your class took a test and:

(a) the SD was 5,
(b) test–retest reliability was + 0·9.

Calculate the SE_{meas} for the test.

$$
\begin{aligned}
SE_{meas} &= 5 \sqrt{(1 - 0\cdot9)} \\
&= 5 \sqrt{0\cdot1} \\
&= 5 \times 0\cdot32 \\
&= 1\cdot6.
\end{aligned}
$$

So if Freddy scored 24 in the test, we can say (because we know that errors tend to follow the normal distribution curve) that his score is 68 per cent certain to be within 1 SE_{meas} or 1·6 points of his true score and 95 per cent certain to be within 2 SE_{meas} or 3·2 points of his true score. Only five times out of 100 would we expect Freddy to get a score more than 3·2 points away from his actual score of 24.

Standard error of the mean

Suppose we measured the weight of every adult male who lived in the City of Durham. We could then report their exact mean weight. Now suppose that George weighed the first 100 adult males to pass him in South Road and John weighed the first 100 adult males to pass him in North Road. Would you expect them to get exactly the same mean weight? Would you expect to estimate *exactly* the same figure for the population mean? However, the bigger the sample in relation to the population, the closer will be the sample and population means. Just as scores fluctuate around the mean score so do sample means fluctuate around the population mean. The measure of fluctuation is called the *standard error of the mean* or SE_{mean}. This is once again simple to compute.

$$SE_{mean} = \frac{SD}{\sqrt{N}}$$

where   SD = standard deviation of sample
and N = size of sample.

Note: There is no SE_{mean} for a population mean.

Supposing John found the following when he weighed his 100 men:
(a) the mean weight was 150 lb,
(b) the SD was 20,
(c) the number weighed was 100.

$$SE_{mean} = \frac{20}{\sqrt{100}}$$
$$= \frac{20}{10}$$
$$= 2 \text{ lb.}$$

John then knows he is 68 per cent certain that his sample mean is within 2 pounds of the population mean and 95 per cent certain that it is within 4 pounds of the population mean. In other words he would be pretty surprised if the population mean turned out to be outside the limits 146 to 154 pounds, and would probably try to find out why his sample differed so widely from the population.

Predicting a Future Score
Standard error of estimation

Finally let us look at the error in predicting a future score from a score we already know on the same test—the *standard error of estimation*. Here the problem is affected by *test sophistication*, which is the effect of getting a better score just because one is more familiar with the test the second time and so can go a little faster, waste less time reading the instructions, and so on. *Regression to the mean* also occurs (here once again there is the effect of chance). If a person scores a long way from the mean then next time he would be likely to score closer to the mean. Thus people scoring very low marks on a test of intelligence are likely to do better (as a group) next time purely by chance. This led early workers with SSN children to claim many things for their methods which were not necessarily true. Similarly people scoring very high marks on an IQ test will not tend to do so well next time.

However, despite these problems, we may predict or estimate a person's future range of scores *on the same test* by using the following formula:

$$SE_{est} = \sigma \sqrt{(1 - r^2)}$$

where σ = standard deviation of scores
             r = reliability figure for test

Taking the example on page 49 again,

$$\sigma = 5,$$
$$r = + 0{\cdot}9,$$
$$SE_{est} = 5 \sqrt{(1 - 0{\cdot}9^2)}$$
$$= 5 \sqrt{(1 - 0{\cdot}81)}$$
$$= 5 \sqrt{0{\cdot}19}$$
$$= 5 \times 0{\cdot}436,$$
$$SE_{est} = 2{\cdot}180.$$

Note this is much larger than the SE_{meas} for the same figures. We are more cautious about saying a future score will fall within a certain range than we are about saying a current score is fairly accurate.

From this we can be fairly certain that a person taking the same tests on another occasion would get a score within 2·18 points of his first score in 68 per cent of the cases, within 4·36 points in 95 per cent of the cases, and so on.

Question 16

Given the following data, calculate:
(a) SE_{mean}
(b) SE_{meas}
(c) SE_{est}
(d) If Charlie took the test again, between what range would you be 95 per cent confident that his marks would fall?

Only half the class turned up for a science test, and these were their marks:

Rita	22	Ned	25
Jean	26	Ethelbert	12
Freda	29	Paul	23
Linda	17	Cliff	19
Norma	20	Charlie	21

The reliability figure for the science test was given as $+ 0.95$.

CHAPTER SIX
Correlations

MANY THINGS HAPPEN in nature which are related to each other—babies eat food and grow in size; new houses mean less land for farming; the sun's setting goes hand in hand with a drop in temperature. It is possible to calculate the amount of relationship between any two sets of happenings or sets of scores. Calculation of such a relationship is called the calculation of the *coefficient of correlation.*

Numerically the results of such a calculation will fall within the limits $+ 1 \cdot 0$ (a perfect positive relation) to $- 1 \cdot 0$ (a perfect negative relation), whatever technique for calculation we care to use. (This is not true for biserial correlations, but we shall not be concerned with these in this book.)

It must also be pointed out that it is possible to get a correlation between two sets of figures *without* there actually being a direct causal relationship. This can happen either from pure chance or because there is a relationship of both factors with another factor. Take the following example:

1. The further up a mountain we travel the less oxygen there is available. If height is plotted against oxygen availability a high negative correlation would be obtained.

2. Oxygen is required for normal life to continue. The less oxygen there is in the air, the more likely we are to suffer from headaches, dizziness, fainting, panting, etc. Again a high negative correlation would be obtained if the incidence of these symptoms was plotted against availability of oxygen.

3. If, however, we plotted the oxygen-lack symptoms against height as we climbed a mountain, we would find that we got a high positive correlation, suggesting that height was responsible for the unpleasant symptoms. However, this can be shown to be untrue by providing adequate oxygen whereupon the symptoms disappear. The two are in fact unrelated but both have their link with a common third, all-important factor. It is therefore important that we remember, that whilst a correlation between two sets of figures *usually* means that

there is a relationship between them, sometimes this cannot be accepted at face value.

Correlations can be shown graphically and the correlation coefficient can also be calculated from a graph. However the process is generally slower than calculation by formula. Also for large samples it involves grouping the data with an inevitable loss of precision. (Calculation from grouped data is covered in Chapter 10.)

Take the following two examples of pupils' form position in arithmetic and English.

1. Name	Arithmetic position	English position	Name	Arithmetic position	English position
Nancy	1	1	Freda	4	4
Bill	2	2	Alan	5	5
Joan	3	3			

If we draw a graph it would look like this:

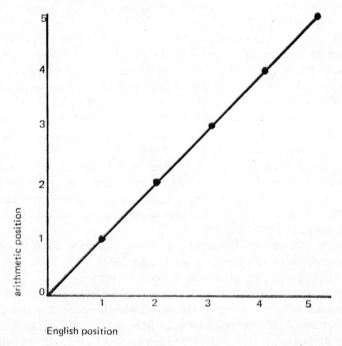

English position

This is a perfect positive correlation of + 1·0.

2.	Name	Arithmetic position	English position	Name	Arithmetic position	English position
	Nancy	1	5	Freda	4	2
	Bill	2	4	Alan	5	1
	Joan	3	3			

Now the graph looks as follows:

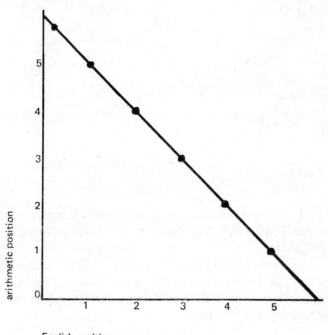

English position

This is a perfect negative correlation of — 1·0.

In fact we rarely get perfect correlations, and scores tend to be scattered away from the lines shown. The more the scores are scattered the lower the correlation will be. High positive, high negative and zero correlation scatter diagrams are shown below. Note their shapes.

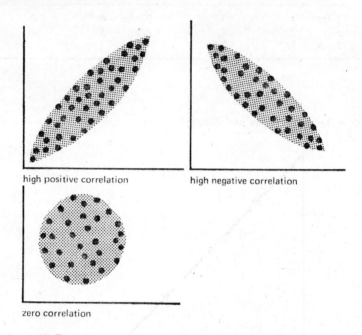

high positive correlation   high negative correlation

zero correlation

A positive correlation 'leans' from left to right, a negative correlation leans from right to left and a zero correlation has not got any direction to it.

Some people find it hard to understand how a negative correlation can tell us just as much as a positive correlation, but perhaps the following illustrations will help. Suppose we have two weights joined by a piece of string.

weight A   weight B

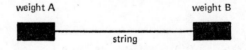

string

Now if we join another piece of string to the middle of the first piece and then pass it over a pulley we can pull the weights up or lower them at will.

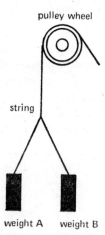

pulley wheel

string

weight A   weight B

The amount that weight A *rises* will correlate perfectly with the amount that weight B *rises*, i.e. a perfect *positive* correlation.

Now suppose we hang our two weights over the pulley *without* the second piece of string.

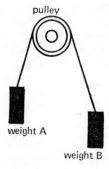

pulley

weight A

weight B

The amount that weight A *rises* will now correlate perfectly with the amount that weight B *falls*, i.e. the correlation between *rise* of A and *rise* of B is perfect and negative. In *both* cases we can easily deduce where B is if we know where A is.

Calculating Correlations
Spearman's rank order correlation

A simple method of calculating a correlation is *Spearman's rank order correlation*. This is useful for classes of children (or any set of scores) up to a maximum of thirty scores in each set. Beyond thirty the results tend to be unreliable.

For this method the first set of scores is arranged in rank order and then the second set of scores is set next to them, e.g.

Name	Arithmetic score	Rank	English score	Rank
Nancy	27	1	47	2
Bill	21	2	58	1
Joan	19	3	12	5
Freda	17	4	35	4
Alan	6	5	38	3

Now we are in a position to use Spearman's formula:

$$r = 1 - \frac{6 \sum d^2}{N(N^2 - 1)}$$

So our example becomes:

Name	Arithmetic rank	English rank	Difference score	d^2
Nancy	1	2	1	1
Bill	2	1	1	1
Joan	3	5	2	4
Freda	4	4	0	0
Alan	5	3	2	4

Therefore
$$\sum d^2 = 10.$$

The size of population or
$$N = 5.$$

So substituting the figures in the formula:

$$r = 1 - \frac{6 \times 10}{5\,(25 - 1)}$$

$$r = 1 - \frac{60}{120}$$

$$= 1 - 0\cdot5,$$
$$r = + 0\cdot5.$$

Correlation is therefore positive and present.

Partial tetrachoric correlation

Another simple method which is useful in the classroom is the *partial tetrachoric*. It is however only a rough guide and depends for its accuracy on the assumption that the scores are scattered uniformly. This is most easily shown diagrammatically:

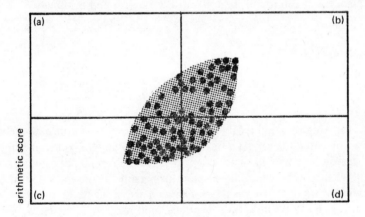

arithmetic score

English score

The graph is divided into four quadrants:
(a) Pupils high for arithmetic but low for English.
(b) Pupils high for arithmetic and also for English.
(c) Pupils low for arithmetic and also for English.
(d) Pupils low for arithmetic but high for English.
The calculation is only based on the quadrant (b) and the assumption is that the shape of the curve in (b) holds good for the rest of the graph.

All we have to do by this method is to put the two sets of scores in rank order. Take the top half of the class for arithmetic and see *how many are also in the top half* for English (or whatever we are comparing). Divide this number q by the population of the whole class, convert to a percentage and look up the correlation in Table 1, page 124.

For example:

Name	Arithmetic rank	English rank	
Nancy	1	2	
Bill	2	1	Top half
Joan	3	5	
Freda	4	4	
Alan	5	3	
George	6	6	

Nancy and Bill are in the top half for English and arithmetic.

$$q = 2,$$
$$N = 6,$$

$$\frac{q}{N} = \frac{2}{6} \times 100 = 33\tfrac{1}{3} \text{ per cent}$$

Looking this up in Table 1 we see that $r = + 0 \cdot 5$ approximately.

You should never get a fraction greater than a half by this method because only half the class can ever be in the top 50 per cent.

When you have an odd number of persons in the class you may wonder what to do with the median person. Is he in the top or bottom half? As he is only a number and not a real person we can divide him. If he is above the median on the second test, count him as half of a person. If he is at the median for both, count him as a quarter of a person and if he is below the median on the second test then obviously do not count him at all.

Because the partial tetrachoric only uses evidence from one of the four quadrants it is not a good method for very small groups. In general, it is quite reasonable for average class-size groups in excess of twenty-five in number, and so can be used in a complementary fashion with Spearman's rank order method covering the smaller group.

Pearson's product moment correlation

Spearman and the partial tetrachoric are both rank order correlations. As such they operate as if scores are spread uniformly across
the range of marks. Product moment correlations are calculated
from the actual marks gained and do not assume that there is an
equivalent distance between each testee and those above and below
him. The one we shall be looking at is widely used; it is called
Pearson's product moment correlation.

At first sight this formula for calculating correlations looks
formidable. However, it is really only a lot of simple steps all rolled
into one formula.

$$r = \frac{\sum xy - \frac{(\sum x)(\sum y)}{N}}{\sqrt{\left\{\left[\sum x^2 - \frac{(\sum x)^2}{N}\right]\left[\sum y^2 - \frac{(\sum y)^2}{N}\right]\right\}}}$$

r is the correlation coefficient, i.e. the thing we wish to find, and
$\sum xy$ is the sum of all the x-scores times all the y-scores for each
person, i.e. multiply each person's scores together and then add
the lot up. $\frac{(\sum x)(\sum y)}{N}$ means add up all the x-scores, add up all
the y-scores, multiply the two together and divide by N, the number
of people in the sample.

$\left[\sum x^2 - \frac{(\sum x)^2}{N}\right]$ means the sum of each x^2 minus the sum of all

the xs which is then squared and divided by N.

$\left[\sum y^2 - \frac{(\sum y)^2}{N}\right]$ means the same as for x, but substituting y.

Multiply these two parts together and take the square root of
the product. Finally we need to divide the numerator by the denominator. The resulting fraction is our figure for r, our correlation
coefficient.

Generally we get most of our figures by constructing a table as
shown in the following example:

Calculate the Pearson r between the arithmetic and English marks
for the following boys.

	Arithmetic	English
Arthur	7	7
Fred	6	5
George	8	3
Norman	4	4
Brian	9	5
Peter	5	6
Albert	4	8

Call the marks for arithmetic x and the marks for English y.

x	y	x^2	y^2	xy
7	7	49	49	49
6	5	36	25	30
8	3	64	9	24
4	4	16	16	16
9	5	81	25	45
5	6	25	36	30
4	8	16	64	32
$\sum x = 43$	$\sum y = 38$	$\sum x^2 = 287$	$\sum y^2 = 224$	$\sum xy = 226$

$N = 7.$

Substitute in the formula:

$$r = \frac{\sum xy - \dfrac{(\sum x)(\sum y)}{N}}{\sqrt{\left\{\left[\sum x^2 - \dfrac{(\sum x)^2}{N}\right]\left[\sum y^2 - \dfrac{(\sum y)^2}{N}\right]\right\}}}$$

$$r = \frac{226 - \dfrac{43 \times 38}{7}}{\sqrt{\left\{\left[287 - \dfrac{43^2}{7}\right]\left[224 - \dfrac{38^2}{7}\right]\right\}}}$$

$$r = \frac{226 - \dfrac{1634}{7}}{\sqrt{\left\{\left[287 - \dfrac{1849}{7}\right]\left[224 - \dfrac{1444}{7}\right]\right\}}} = \frac{-7 \cdot 4}{\sqrt{(23 \times 18)}}$$

$$r = \frac{-7 \cdot 4}{\sqrt{414}} = \frac{-7 \cdot 4}{20 \cdot 35} = -0 \cdot 364.$$

Question 17

Calculate the Spearman rank order correlation for the previous example, using the same scores for each boy.

What does *r* measure? As was mentioned in Chapter 4, a great deal depends on whether *r* is measuring reliability or validity. Generally the majority of calculations which we do are likely to belong to the second category. In this case, *r* is able to tell us many things:

1. Where the correlation between two tests is very high (above 0·80 or 0·85) then we don't really learn much more about a person by giving him the two tests—they may well be measuring much the same things (even if they have very different titles).

2. When the correlation between two tests is very low (below 0·20 or 0·15) then it is worth giving the two tests because they are probably measuring different things about a person.

3. If we are using a test as a predictor of future performance then the results need to correlate fairly highly with the future performance if we are not to make many faulty decisions.

4. A negative correlation can tell us just as much about the things we are testing as can a positive one.

For instance, heart attacks correlate with certain things in men as follows:

(a) Smoking against likelihood of heart attack—positive *r*

(b) Age against likelihood of heart attack—positive *r*

(c) Height against likelihood of heart attack—negative *r*

(d) Regularity of exercise against likelihood of heart attack—negative *r*

What does this mean?

(a) The older men get the more likely they are to suffer a heart attack; (b) the more heavily they smoke the more likely they are to suffer a heart attack,* (c) the shorter they are the more likely they are to suffer a heart attack; (d) the longer the gaps between regular exercise the more likely they are to suffer a heart attack. Putting it bluntly a short, middle-aged, heavy smoker who digs the whole garden in one Sunday afternoon after spending the whole winter doing nothing is asking for trouble.

Lastly in this chapter it is appropriate to introduce two new concepts: parametric and non-parametric.

Parametric tests are those which use the actual scores as the basic source of data for calculations. Non-parametric tests are less powerful. They are tests where the calculation is based on rank orders or the number of times things occur in certain categories (called frequency). Thus Pearson's method is a parametric method whilst Spearman's and the partial tetrachoric are non-parametric. As a rule we should use the most powerful test available *but* not if this means we are using a parametric test although we only have non-parametric data.

---

* In the light of some recent evidence the validity of these statements is now less certain. If the following piece of research can be substantiated the relationship would appear to be *non-causal*.

Cederlöf and others have studied tens of thousands of identical twins. (Identical twins are believed to be genetically identical.) Their interest centered on pairs in which one twin smoked and the other did not. They found the smoker was no more prone to heart disease than the non-smoker! It may be that whatever it is that increases the likelihood of heart disease can also increase the need to develop a relaxing habit such as smoking.

CHAPTER SEVEN
Significance

IF TEN SCHOOLS are in a football league and school A beats all of the other schools whilst school B loses to all of the other schools including school A we should not have difficulty in deciding that A was superior to B at football. However, if school A and school C both beat the eight other schools decisively but failed to play each other because of bad weather then we would have a much more difficult task deciding which was the better team. For the benefit of a league it would probably be decided on goal average but many supporters of the team coming out second would undoubtedly say, 'Ah yes, but it would have been different if we'd played the other team.' By subjecting the two sets of goals to a test of significance it would be possible to see if the difference was in fact a real one—in so far as goals are concerned that is.

Let us take another case. When you toss a penny up in the air would you expect it to come down roughly the same number of heads as tails? The answer is yes. If it does not work out like that you would start to watch whoever was tossing the coin *or* perhaps ask to have a closer look at the coin. In other words, common experience has taught you to suspect that pure chance isn't all that is affecting the coin. You would start to wonder if the difference was *significant*.

There are several ways of calculating whether a difference is significant or not. When we decide whether a difference is merely due to chance on that occasion or could happen again, we state how confident we are that the difference is a real one. *Significance levels* commonly used are:

5 times out of 100   once out of 100   once out of 1000

In effect what we are saying is, 'That difference could only happen five times in a hundred by chance alone' (or once in one hundred times or once in one thousand times). These figures are reported as follows:

5 times out of 100   as 0·05   level of significance
once   out of 100   as 0·01   level of significance
once   out of 1000 as 0·001 level of significance

Here we shall look at three ways of calculating significance and one of measuring significance of correlations.

1. The *t*-test of significance for small populations.

2. Test for significance difference between the means, for larger populations.

3. The *F*-ratio test of significance for small populations (see page 87).

4. Significance of size of correlations.

Before that, however, take a look at the following two sets of scores:

A	
7	$N = 23,$
10	$M = \dfrac{\sum A}{N} = \dfrac{265}{23}$
9	
10	Mean of A scores $= 11\cdot52.$
11	
13	**B**
11	7
15	10
11	8
11	11
8	9
12	–
10	$\sum B = 45,$
12	$N = 5,$
12	$M = \dfrac{\sum B}{N} = \dfrac{45}{5}$
12	
14	Mean of B scores $= 9.$
12	
16	
13	
9	
13	
14	

$$\sum A = 265,$$

Are these two sets of scores significantly different? At first glance we might be tempted to say yes. However, let us first look at them on the same graph.

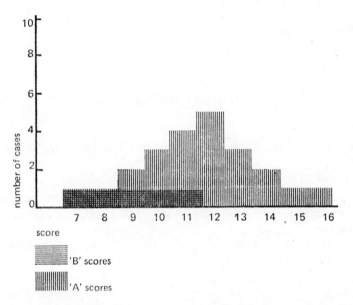

'B' scores

'A' scores

Here B scores can be seen to be a sample of scores which fall within the spread of the A scores. The different mean scores alone do not show this. However, the sigma for A would be much larger than that for B and so would lead us to suspect that perhaps the difference was significant. (A one-tailed test might well produce different results from the methods described below. More advanced texts should be consulted about this.)

The t-Test of Significance

When we are dealing with small samples (below thirty) there are usually not enough scores to give an approximation to the normal curve. As a result a man called Gosset, writing under the pseudonym Student, developed a test called the t-test of significance. By using it we are able to estimate the standard error of the difference between the means of two samples or two small populations (t-test can also be used for large samples or populations. The reverse is not always true for other formulae.)

In order to use the formula, one new concept needs to be introduced—*degrees of freedom*. Briefly, this is numerically equal to the number of changes which we can make to a group of figures (scores, etc.) whilst still satisfying any external requirements. For instance, suppose we have four numbers which must add up to a total of 11. The first three numbers which we choose can be any at all *but* the fourth number is restricted (it is not free) for it must ensure that the total comes to 11. For example:

2 3 1 $\boxed{5}$ = 11 The number in the box can only be 5.

9 4 6 $\boxed{-8}$ = 11 Here the number in the box can only be — 8.

This means that of the four numbers in each sequence three are free and one is fixed by the others. There are three degrees of freedom.

When we are dealing with samples there is usually one less degree of freedom than the size of the sample, i.e. degrees of freedom = N — 1. This is not true for all calculations, but it is true when we are comparing two small groups in order to see if the difference between them is significant. The formula we use is as follows:

$$t = \frac{M_1 - M_2}{\sqrt{\left[\left(\frac{\sum d_1^2 + \sum d_2^2}{N_1 + N_2 - 2}\right)\left(\frac{1}{N_1} + \frac{1}{N_2}\right)\right]}}$$

where M_1 = mean of scores for first group,
M_2 = mean of scores for second group,
$\sum d_1^2$ = sum of squared deviations from the mean for the first group,
$\sum d_2^2$ = sum of squared deviations from the mean for the second group,
N_1 = number of scores in first group,
N_2 = number of scores in second group.

Choose the larger mean to be M_1. This avoids having a minus sign in the numerator.

Degrees of freedom = $(N_1 + N_2 - 2)$ (— 2 because there are two samples).

As a result of working out this calculation you will finish up with a number. Look up Table 2 on page 125. Against the degrees of freedom for your particular problem there are two numbers. If your number is smaller than either of them then the difference is not significant. If it is as large as or larger than the first but not as large as the second it is significant at the 0·05 level. If it is as large as or larger than the second then it is significant at the 0·01 level.

Compare the following two sets of arithmetic scores for class 1B and class 1A. Is the difference between them significant?

1A	1B
9	7
7	4
6	6
9	3
5	7
8	8
4	5
7	1
9	4
6	5

$\sum x = 70,$ $\sum y = 50,$
$N = 10,$ $N = 10,$
$M_1 = 7,$ $M_2 = 5,$
Degrees of freedom $= 9.$ Degrees of freedom $= 9.$

For this example I have rounded the means off to whole numbers.

x scores	d_1	d_1^2	y scores	d_2	d_2^2
9	2	4	7	2	4
7	0	0	4	1	1
6	1	1	6	1	1
9	2	4	3	2	4
5	2	4	7	2	4
8	1	1	8	3	9
4	3	9	5	0	0
7	0	0	1	4	16
9	2	4	4	1	1
6	1	1	5	0	0

$$\sum d_1^2 = 28,$$
$$\sum d_2^2 = 40.$$

$$t = \frac{M_1 - M_2}{\sqrt{\left[\left(\dfrac{\sum d_1^2 + \sum d_2^2}{N_1 + N_2 - 2}\right)\left(\dfrac{1}{N_1} + \dfrac{1}{N_2}\right)\right]}}$$

$$t = \frac{7 - 5}{\sqrt{\left[\left(\dfrac{28 + 40}{10 + 10 - 2}\right)\left(\dfrac{1}{10} + \dfrac{1}{10}\right)\right]}}$$

$$t = \frac{2}{\sqrt{\left(\dfrac{68}{18} \times \dfrac{2}{10}\right)}} = \frac{2}{\sqrt{\left(\dfrac{136}{180}\right)}} = \frac{2}{\sqrt{0.7556}}$$

$$t = \frac{2}{0.8692}$$

$$t = 2.301.$$

Look up degrees of freedom, $(10 + 10 - 2 =)$ 18, in Table 2. Our t is larger than that required for the 0·05 confidence level (2·101) but smaller than that required for the 0·01 confidence level (2·878). Therefore, we can say that such a difference between the two classes could only occur five times out of 100 by chance.

The t-test should only be used when certain conditions are fulfilled, these are:

1. The scores are drawn from a population which is reasonably normal.

2. The two samples being compared are roughly the same size.

3. The two samples are not too small (five in each is usually regarded as the minimum.)

4. The two samples have roughly the same variance (Standard deviation).

5. The two samples should be uncorrelated. In essence this means the standard form of t-test should not be used for such things as comparing pre and post test results. The same people having taken the two tests makes for a relationship between them *even* if this relationship is zero!

Tukey's Test

This is a delightfully improbable non-parametric test of significance. The two sets of scores are compared by simply placing them alongside each other and counting up how many scores in one set are below the bottom score in the other set and adding this to the number of scores in the second set which are above the top score in the first set.

Example:

Scores group (a) 3,4,4,5,5 5 6,6,7,7,8,9.
Scores group (b) 6,7,7,7,8,8,9,9 10,11,11,12.
Group (a) scores below the lowest group (b) score total 5.
Group (b) scores above the highest group (a) score total 4.
$4 + 5 = 9.$

In Tukey's test there are only three critical ratios to remember: 7 at the 0·05 level; 10 at the 0·01 level; 14 at the 0·001 level of significance.

So in the above example the difference between the two groups is significant beyond the 0·05 level.

Significance Levels for Large Samples

When we have larger samples to compare, we no longer need to worry about the degrees of freedom involved, because the change in the value of the *critical ratio* (the number we calculate) is small enough not to matter at the 0·05 and 0·01 levels of significance. Indeed we only need to remember two approximate numbers:

1. If CR (critical ratio) = more than 1·96 then significance is at or greater than the 0·05 level.
2. If CR = more than 2·58 then significance is at or greater than the 0·01 level.

The formula we use is as follows:

$$CR = \frac{M_2 - M_1}{\sqrt{[(SE_{mean_1})^2 + (SE_{mean_2})^2]}}$$

where M_2 = the mean of second sample
and M_1 = the mean of first sample.

(Make the group with the larger mean your second group.)

SE_{mean_1} = standard error of mean of group 1,
SE_{mean_2} = standard error of mean of group 2.

The standard error of the mean is calculated by using the formula given in Chapter 5.

$$SE_{mean_1} = \frac{SD_1}{\sqrt{N_1}}$$

I would suggest that rather than substituting $\frac{SD}{\sqrt{N}}$ in the first formula you calculate the two SE_{means} and substitute the actual values in the formula. This avoids the confusion of having a pair of fractions within another fraction.

Compare the eleven plus performance of the following two groups of children from two schools. Is the difference significant?

School A		*School B*	
Number of children	81	Number of children	64
Mean score	123	Mean score	113
SD	10	SD	14

Firstly compute the standard errors of the means.

$$SE_{mean_2} = \frac{10}{\sqrt{81}} \qquad\qquad SE_{mean_1} = \frac{14}{\sqrt{64}}$$
$$= 1 \cdot 111, \qquad\qquad\qquad = 1 \cdot 75.$$

Now substitute in the formula.

$$CR = \frac{M_2 - M_1}{\sqrt{(1 \cdot 75^2 + 1 \cdot 111^2)}} = \frac{123 - 113}{\sqrt{(3 \cdot 063 + 1 \cdot 234)}}$$

$$CR = \frac{10}{\sqrt{4 \cdot 297}} = 4 \cdot 8.$$

Therefore significance is beyond the 0·01 level. In other words such a difference would occur less than once in one hundred times by chance.

It is worth pointing out at this stage that whilst we are able to use statistical means to find out whether a significant difference exists between two groups they will not tell us why; for instance, in the last example we do not know whether the difference is due to:

(a) a difference in teaching methods,
(b) a difference in children's ability,
(c) the results of coaching in only one school,
(d) an error in examining techniques at one school, etc.

Significance of Correlation Coefficient

Suppose two boys were to each toss two pennies into the air and the pennies all came down heads. The correlation between the first boy's throw and the second boy's throw would be perfect, i.e. a correlation of $+ 1 \cdot 0$. However, common experience tells us that this is a fairly likely happening once in a while if only pure chance is operating (once in sixteen times in this case). If the two boys were each to toss twelve coins into the air we would be very surprised if they came down showing twenty-four heads. From this simple illustration I hope you can see that it is possible to get a high correlation which is not necessarily significant if the number of cases in the sample or population is very small. The larger our population the smaller r will have to be in order to be significant.

Take for example the populations $N = 10$, $N = 100$ and $N = 1000$. For significance at the 0·01 level the correlations need to be 0·765, 0·256 and 0·081. Table 3 on page 126 gives the values of r required for significance at the 0·05 and 0·01 levels.

There are other methods of determining significance. One such test is the χ^2 test (called chi-square, pronounced ki). This test involves making an hypothesis about a group of people (or whatever we are measuring). For example, 'I predict that all of the men under 5 feet 6 inches tall will be afraid to dive from the 10-metre board'. We then test the hypothesis and the degree to which it is untrue is reflected in the size of the χ^2 ratio that we obtain.

χ^2 is described in Chapter 12. The methods already mentioned should however be sufficient for most elementary calculations of significance and indeed for many used in educational research. E. F. Lindquist points out that the χ^2 distribution is generally considered to be one of the most important in statistical theory.

Question 18

The boys and girls in the eleven-year-old class were asked to run 100 yards as fast as they could. From the following results calculate whether the difference between the boys and girls was significant.

Name	Time	Name	Time
Magnus	13·4	Jean	14·7
Malcolm	13·9	Joan	14·0
Manuel	14·6	Jill	13·9
Mark	15·2	Joyce	15·0
Marmaduke	14·1	Jane	15·1
Martin	15·0	Janet	13·7
Matthew	13·2	Joanna	13·9
Morris	13·9	Josephine	14·0
Max	14·8	Jacqueline	14·2
Michael	14·9	Jocelyn	14·3
Miles	15·7	June	14·2
Monty	14·6	Judith	14·6
Mortimer	14·5	Jessica	14·5
Moses	14·4	Jenny	14·3
Mungo	14·0	Jemima	14·7

You will need to calculate several things for this problem. Firstly you will need to decide which formula you will use. Remember that a low time means a fast runner not a slow one.

If we take an example of the t-test actually being used in a piece of research the work reported in *Educational Research* of November 1965 is of interest. W. H. King compared mathematical achievement in London, in large and small suburban schools. During the survey many children failed to finish one or more tests.

He found that the mean scores of those who completed all three tests were superior to the mean scores of those who failed to complete a paper. Part of his results are given below. Decide the level of significance of the difference for the three separate tests.

Means for Complete and Incomplete Test Scores

Group	Raven's PM N	M	Step 3A (Maths) N	M	Maths I N	M
Complete	3134	47·30	3134	30·69	3134	69·99
Incomplete	117	44·13	132	24·20	74	53·52
t	—	4·61	—	7·44	—	6·19

(To avoid any confusion, it should be pointed out that the means against the row labelled 'incomplete' are the means for those children who completed *that particular paper*, but who failed to complete the full range of tests.)

CHAPTER EIGHT
Objective testing and item analysis

THERE ARE MANY ways of testing knowledge in the classroom or during some piece of research. Objective testing, by reason of the speed of marking and high level of consistency *whoever* does the marking, is slowly gaining ground in this country. It has been a popular way of testing in the United States of America for many years. There are many arguments both for and against objective testing just as there are for and against essay testing. Of course we tend to defend the system with which we have grown up.

Of necessity in a small introductory statistics book for non-mathematicians the amount of space given to a discussion of objective testing must be brief. I hope the following few paragraphs will help those not already familiar with this topic.

Advantages
1. As mentioned before, with a truly objective test it doesn't matter who marks it or when it is marked, it will get the same score. Two sources of error are (a) the possibility of miscounting the number of correct responses and (b) of cutting a marking 'key' in the wrong place for one or more questions. These errors are minute when compared with the variations in essay marking. Here well-known distortions are caused by some examiners refusing to use the full range of marks, becoming bored as the task progresses, lowering standards as the task progresses, being affected by their own emotional state of 'well-being' and also being affected by their own levels of knowledge.

2. They are very quick to mark compared with essays.

3. It is relatively easy to check the 'goodness' of individual questions via item analysis.

4. It is relatively easy to place questions in an accurate order of difficulty via information obtained from item analysis.

5. To my mind the biggest advantage is that for a given amount of time given over to testing it is possible to sample much more of the taught material than via traditional essays. This means examinees

have to read/revise the whole range of material rather than opt for a few 'banker' topics. There is also less luck involved in getting a high mark 'because the right questions came up'.

6. Result feedback to examinees can be very quick.

Frequently Mentioned Disadvantages

1. Objective tests encourage the cramming of facts rather than an understanding of, and ability to use, knowledge. This is often true, not because it needs to be but because 'factual' questions are easier to construct. Whilst it is no answer to say the same often applies to essay answers, nevertheless I seem to have marked many hundreds of essays from GCE through to degree finals where the question encouraged this same behaviour.

2. It takes a very long time to write a series of good objective questions. This does tend to balance out the gain on marking time.

3. Many argue that objective tests encourage guessing. Certainly if left purely to the examinee's choice then the gambler would have an advantage. The ways of correcting for guessing are mentioned at the end of this chapter.

4. Objective questions tend to demand totally right or totally wrong responses with no way of saying 'I agree, but . . .'. Similarly there is usually no way of giving marks for nearly-right answers.

5. Objective tests are often criticized because they do not allow the examinee to express himself freely or to develop particular themes. There is no answer to this. They don't. However, the objective test is merely one tool in a whole range of evaluative instruments which we as teachers should have at our finger tips. Different methods of testing are appropriate for different kinds of knowledge and skills. If we want to know how many strawberries a boy has picked, we don't set an essay. Part of the teacher's task is to choose appropriate testing techniques which will stimulate the child to acquire knowledge *and* best measure how much he has acquired. With that in mind I hope the remainder of this chapter will help those who wish to develop, evaluate and mark objective questions.

Types of Objective Question

There are various types of objective item, (the word 'item' is used in objective testing jargon synonymously with the word 'question'); some are more objective than others.

1. Gap sentences where a word is, or words are, left out e.g. 'Too many . . . . . spoil the broth'.

2. Open ended sentences where the examinee provides a final word or words. e.g. 'Too many cooks spoil the . . . . .'

With both of these types a common construction error is to write questions with too many gaps or where a variety of words make for a sensible answer. It is *always* wrong to argue 'they should have known what I wanted!'
e.g. 'Too many . . . spoil the . . . .'
How about 'Too many sweets spoil the appetite'
or even 'Too many tests spoil the desire to learn for learning's sake'.

3. True/False or Right/Wrong questions. Typically a statement is made and the examinee has to decide whether it is correct or not.
  e.g. 'Nixon followed Kennedy as the President of the United
        States of America'.
        Right,
        Wrong.
Here we have a good example of an ambiguous question. Does it mean 'immediately after', or 'after' rather than 'before'? A more intricate, some would say tricky, version of the true/false question is where two statements are made both of which have to be correct for the question to be answered 'true'.
  e.g. 'Winston Churchill, who died at age ninety-five, inspired the
        British people during the Second World War'.
Whilst most would agree with the second part, the first being wrong means the whole statement has to be regarded as false.

4. Multiple Choice. In this type of question a feeder, called the stem, leads into a variety of possible endings.
  e.g. Deciduous Trees

    (a) Grow taller than conifers   (a) ☐

    (b) Lose their leaves in Winter   (b) ☐

    (c) Need to have their roots in running water   (c) ☐

    (d) Only grow in the Northern Hemisphere   (d) ☐

It is easier to write short wrong choices than short correct choices. This needs care as it can 'give away' the most likely answer to someone who doesn't know. All responses should be equally plausible to someone who lacks the specific information, otherwise he can arrive at the right response by elimination.

5. Matching questions. Here typically two lists are given and the examinee decides which ones make up to form a pair. One list should always be longer than the other. In this way an examinee cannot get the last answer by the process of elimination rather than knowledge.

e.g. Match the following towns with the countries to which they belong.

Towns	Countries		
1. London	(a) Eire		
2. Swansea	(b) Northern Ireland	1	
3. Glasgow	(c) Scotland	2	
4. Cork	(d) Scillies	3	
	(e) Wales	4	
	(f) England		

General Tips

1. Because many objective questions depend upon the *recognition* of learned material rather than *recall* from memory, you can typically expect higher marks from this type of test than from essays. Don't be misled into thinking knowledge is necessarily greater as well.

2. The sensible teaching practice of going from the known to the unknown applies just as much to testing techniques as to knowledge being learned. It will reduce unnecessary worry on the part of your pupils if they are given a short practice test some time prior to the real thing. Once they are familiar with this type of test, the practice can be abandoned.

3. Key words in objective tests are often words such as 'may' and 'will', 'always' and 'usually', 'sometimes' and 'occasionally', etc.

(a) 'If you throw rocks at policemen, and hit them, you *may* kill them.'

(b) 'If you throw rocks at policemen, and hit them, you *will* kill them.'
are two very different statements. Children (and adults very often!) need to be shown how to read carefully. Objective tests aren't meant to be tests of reading ability but they do carry the fringe benefit of encouraging accuracy of word useage.

4. It is possible to have either a strict time limit or no time limit at all (unlike essays this cannot lead to the 500 page short essay!) I never set time limits to my tests as I feel this allows the different temperaments of examinees fuller scope and doesn't involve contamination of results because of differing reading speeds. Over the years I've found that very early and very late finishers *tend* to do less well than the rest.

5. When setting out your test paper avoid a pattern for correct answers creeping in e.g.
question   1 2 3 4, 5 6 7 8, 9 10 11 12, 13 14 15 16, 17 18 19 20.
correct answer   a b c d, b b c d, c b c d, d b c d, a b c d.
The wily examinee will spot it. Worse, other examinees *may* become unwittingly trained by you to produce the pattern (called response set) and to continue to do so even after you have abandoned the pattern.

6. Jot down possible questions whilst preparing the lesson, reading the text book or writing notes for hand-outs. It is much harder to think of enough questions at the end of a course of lessons.

7. A common failing is to assume that the question as written, or the chosen answer, is good. Never write questions and use them without operating a vetting system. Either put them into a drawer for a couple of weeks and then re-read them or get a colleague to try to tear them to pieces (this hurts, it's like watching your teddy-bear enter the jaws of a mad dog.) Preferably both of these processes should be followed because so many questions emerge as being ambiguous, lacking essential information, ungrammatical, or even demanding the wrong response.

8. Check that you haven't eliminated certain multiple choice responses by changing from singular in the stem to plural in the responses, or vice versa. Similarly you should check for tense.

9. It is generally agreed that questions which are negative statements should be avoided. Double negatives should never be used.

10. Put potential answers which have an alphabetical or numerical order in ascending (or descending, it doesn't matter) order.

e.g. A kilogramme is

(a)   10 grams

(b)   100 grams

(c)   1000 grams

(d) 10,000 grams

(Now look at the example on matching questions. How should it have been written?)

Having written a test, it is necessary to see whether all of the questions should be retained in future versions of the test, rewritten, dropped altogether, perhaps put earlier or later in the test. Such a process is called *Item Analysis* and is commonly broken into two parts, item discrimination and item difficulty.

1. Item Discrimination

If we are using a test to select certain people and reject others, we need to know whether the individual questions are doing a good job of work. The method of doing this is to compare those people who do well on the whole test with those who do poorly on the whole test. The recommended portions are the top and bottom $27\frac{1}{2}$ per cent of the tested population. However, this is not a figure to which we have to stick rigidly, or indeed *can* always use. In practice anywhere between the top and bottom 25 per cent or top and bottom $33\frac{1}{3}$ per cent is quite adequate. If you have a fairly small population it may well be best to split it in half and compare the top and bottom 50 per cent.

Having sorted out two samples (let us say the top third and bottom third), we then compare the scores of these sub-groups on each question, or item as it is called. We would expect more of the top third to get it right than the bottom third in each case—if the question is doing its job that is. The formula we use is a simple one. (This formula can only be used for questions where the answer is clearly right or wrong.)

$$D = \frac{H - L}{N}$$

where H = sum of scores of high group on a particular question,
$\quad L$ = sum of scores of low group on the same question
and $\quad N$ = number of pupils in one of the groups (they should both be the same size).

The results that we get will fall somewhere between -1 through 0 to $+1$. A question which discriminates perfectly will get a score of $+1$. A question which does not discriminate at all (i.e. the same number in both groups got it right) will get a score of 0. A question which favours the low group will produce a negative result.

(a) Items with a negative result should *never* be used again.

(b) Items with a positive discrimination index below $+0.3$ are of doubtful value for prediction. (However, in class tests we often set easy questions to start with in order to ease the children into the test, so everyone manages to get them right. In other tests we hope that everyone will get all the questions right. Therefore the reason for the test must help us to decide whether to include an item again.)

George, Bill, Geoff, Philip and Reg were the top five boys out of fifteen for a chemistry test. Harry, Neville, Sidney, Claude and Clarence were the bottom five boys in the same chemistry test. Compute the item discrimination index for question seven if they got the following results:

High group		Low group	
Name	*Right or wrong*	*Name*	*Right or wrong*
George	√	Harry	×
Bill	×	Neville	×
Geoff	√	Sidney	×
Philip	√	Claude	×
Reg	×	Clarence	√
$H = 3$		$L = 1$	

$$\frac{H - L}{N} = \frac{3 - 1}{5} = \frac{2}{5}$$

Item discrimination index $= +0.4$.

Therefore the question is doing a reasonable job of work. In practice of course we would not draw up a table. Instead we would just count up the rights in the high group and jot down the number then repeat for the low group.

2. Item Difficulty

By the end of a test we may want a few questions which stretch even the most able pupils. Experience helps us to make this decision. A calculation of difficulty can confirm or contradict our choice of question order. Item difficulty is calculated simply by using the following formula:

$$\frac{\text{Right}}{N} \times 100\% = \text{"difficulty"}$$

where Right is the number who get the item right in the whole class, and N is the number of children in the *whole* class.

The more people that get a question right, the higher will be the question's *difficulty index*. An odd convention, but one that is generally used.

It is reasonable policy for the majority of the questions to have a difficulty index falling between 40 per cent and 60 per cent. (Note that a high percentage means an easy question.) When the difficulty index is measured, the children getting the question right would, of course, be expected to be in the top part of the class. This means that we should normally calculate both the difficulty and discrimination indices for each question. Some authorities say a good question is one of difficulty around 50 per cent and discrimination index above + 0·33 *if* we want to get our children into a merit rank order as a result of the scores obtained on that test.

Just as we may want a few questions to stretch the most able pupils, so it is often desirable for everyone to be able to answer the early questions fairly easily. Item difficulty indexes can help us to decide the total order of the questions in a test from the easiest to the most difficult.

Guessing Corrections

Two different approaches are possible to help us reduce the effects of guessing in objective tests. One is *to instruct examinees clearly that they must respond to every question*, making it plain that by so doing they aren't penalized when compared with the more willing gamblers.

It should be easy to see from simple probability that given four potential answers the wild guesser will *guess correctly once every four times. OR,* putting it another way, given four choices of answer per question *then for every three questions guessed wrongly one will be guessed* correctly. (As with all variables this too is subject to normal distribution error. Some examinees will be more lucky, others less lucky, than average when they guess.)

The alternative of applying a guessing correction depends upon this probability. If a guessing correction is applied examinees *must be instructed clearly "Not to guess."*

A simple formula enables us to correct scores for guessing:

$$\text{Score} = R - \left(\frac{W}{N-1} \right)$$

where Score = Final mark after correction for guessing.

R = Raw score of 'right' responses.

W = Number of 'wrong' responses.

N = Number of responses following the stem.

Unanswered questions are not included in the calculation.

Example

Suppose in a test of 20 questions 18 were answered, 12 correctly, 6 wrongly. The questions all had four possible responses.

$$\text{Score} = 12 - \left[\frac{6}{(4-1)} \right]$$
$$= 12 - 2$$
$$= 10.$$

If you intend to use a guessing correction it is important that you do not have too many different response sizes. If some questions have two, some three, some four and some five responses following the stem you will have four guessing calculations to make *for every examinee.* Also by virtue of only having a small number of questions in each of these groups the potential error in the figure calculated is very large.

When I started to use objective tests in the mid-1960s I always used guessing corrections. I found this caused my students much distress. They were very unhappy to get a score of say 31 out of 50, when they knew they had got 35 questions right. According to Macintosh, guessing corrections do very little to alter the rank order of examinees. It merely lowers all of their scores somewhat. To me it seems pretty valueless to do several hours work and finish effectively in the same place as I started having caused unhappiness on the way.

Diamond and Evans in a 1973 article 'The Correction for Guessing' review a series of researches on this topic where by and large the consensus of evidence is that using or not using guessing corrections has very little effect upon reliability or validity. By using guessing corrections we tend to introduce the contaminating variable of examinee personality and by carrying out many arithmetical calculations we increase the possibility of making errors in arriving at individual scores.

Whether you decide to make a guessing correction or not must be a personal decision. Which ever you decide your examinees must know; *and* they must also know what is the appropriate action for them to take.

Appendix 4 is a brief example of an objective test in statistics. It contains a question paper and an answer sheet. There is also an answer key, shaded where *I think* the correct responses should lie. By cutting out the 'correct' boxes on a blank answer sheet, marking becomes very rapid. The key is placed over the examinee's answer sheet and the correct responses show through. You should always take a quick look at the answer sheet to make sure each question has received only one tick!

CHAPTER NINE
Standard scores

RECENTLY I WAS talking to the head of the English department of a grammar school. We were talking about the occasional boy or girl who gets 95–100 per cent in A-level mathematics. She happened to say, 'Of course no one could get marks like that for English.' Why is this? Most English teachers will say, 'Because you can never get a perfect answer in half an hour or whatever time is allowed.' This attitude, by no means confined to teachers of English, is both thoughtless and ridiculous. As examiners they should not be asking, 'What is a perfect answer?' Instead they should ask, 'What is the best answer possible from a child of this age with such and such background experience, in the time allowed?' That answer is then worth maximum marks. At the other end of the scale many teachers rarely give nought for an answer. 'If a person sets pen to paper it is worth something.' Here, of course, they are making the invalid assumption that writing words is equivalent to answering the question. As a result some teachers of arts subjects tend to use a contracted range of marks whilst believing that they are using the full range.

How can we compare marks in physics with marks in history? If John is top for English and Peter is top for mathematics whilst their marks are equal for everything else, which one is top of the form? How can we balance the fact that tutor C gives teaching practice marks from fail up to average and never apparently sees an above average student teacher whilst tutor D only ever sees average student teachers, tutor E apparently sees student teaching which ranges from average to superior and only tutor F apparently sees students whose teaching ability covers the whole range from fail to superior? The answer to all these questions is to use *standardized scores*.

All of the following methods of standardizing scores assume the scores should really have followed the pattern of the normal distribution if they had been marked objectively.

A quick, simplified method is to find the mean score for the whole population and the mean score for each subgroup. If the subgroup mean is lower than the population mean then the difference or deviation is added on to each score within the subgroup.

Mean population—mean subgroup = deviation of subgroup mean.
Revised score = raw score + deviation between means.

Conversely if the subgroup mean is higher than the population mean the difference between the means would be taken off each raw score.

This is rough justice and takes no account of how the scores are spread out. In other words it takes no notice of the size of the standard deviation of each subgroup. One way of getting around this is to convert all scores to standard deviation scores called z-scores (see page 23). This means that an average score is reported as zero and the top and bottom marks will only rarely be greater than $+ 3$ or $- 3$ (99·7 per cent of the population is covered by this range of standard deviations). z-scores, however, have the tremendous advantage over raw scores of having the same mean and standard deviation for every test, irrespective of the original subgroup means and standard deviation.

Suppose two subgroups have the following scores, means and standard deviations given by two different examiners.

Subgroup 1 scores	Subgroup 2 scores
2	2
3	5
4	6
5	7
6	10
$Mean_1 = 4,$	$Mean_2 = 6,$
$SD_1 = 1·58.$	$SD_2 = 2·92.$

Is a score of 10 in group 2 better than a score of 6 in group 1? Change both to standard deviations from the subgroup means. This is done simply by dividing the raw score deviation by the standard deviation, as shown opposite.

In this case the score of 10 is better than the score of 6. Most teachers and parents however would find difficulty in interpreting the fact that Johnny was top with a score of + 1·37 or that Freddy was average with a score of 0.

Subgroup 1
raw score = 6
deviation from mean = + 2

$$\frac{\text{Dev.}}{\text{SD}} = z\text{-score}$$

$$\frac{2}{1\cdot58} = z\text{-score}$$

z-score = + 1·27.

Subgroup 2
raw score = 10
deviation from mean = + 4

$$\frac{\text{Dev.}}{\text{SD}} = z\text{-score}$$

$$\frac{4}{2\cdot92} = z\text{-score}$$

z-score = + 1·37.

As a result of this problem several people have evolved systems of *standard scores*. These are simply methods of reporting a score not in terms of the standard deviation but in terms of a set of scores which always have the same mean and standard deviation. This mean may be 50 or 100 or any number we choose. Similarly the standard deviation may be 10 or 15 or anything we choose. One of the most commonly used standard scores is called the *McCall T-score*; it has a mean of 50 and a standard deviation of 10. Thus if a child's raw score is one sigma below the population mean then his *T*-score would be 50 − 10 = 40. To calculate *T*-scores from the *z*-scores proceed as follows:

$$T\text{-score} = 50 \pm (z\text{-score} \times 10).$$

The ± sign merely tells us that the piece in the brackets should be added to the 50 if the *z*-score is positive (i.e. raw score was above the mean) and taken away if the *z*-score was negative (i.e. raw score was below the mean).

In the previous example the two *T*-scores would therefore be:

$$50 + (1\cdot27 \times 10) = 62\cdot7,$$
$$50 + (1\cdot37 \times 10) = 63\cdot7.$$

In practice, of course, if we require McCall *T*-scores we would calculate them directly from the raw scores by using the following formula:

$$T\text{-score} = 50 \pm 10 \times \frac{\text{dev. from raw mean}}{\text{raw SD}}$$

So in the examples we used before:

$$T\text{-score}_1 = 50 \pm 10 \times \frac{2}{1\cdot58}$$

$$T\text{-score}_2 = 50 \pm 10 \times \frac{4}{2\cdot92}$$

Question 19

Calculate the McCall T-scores for the boys in the example sub-group 1 who got raw scores of 2 and 4. Calculate the McCall T-score for the boy in the example subgroup 2 who got a raw score of 7.

One disadvantage of McCall T-scores is that the range of scores is usually between 20 and 80; only three times out of 1000 will scores fall outside these limits, but parents are used to a scale running from 0 to 100.

CHAPTER TEN

Multiple correlation and partial correlation

IN CHAPTER 6 we looked at correlations between two sets of figures. This type of correlation is called a *zero order* correlation. Often, however, we wish to predict something and have more than one piece of information to help us in the prediction process. For instance we know that successful high jumpers tend to be tall and slim (i.e. lightweight), we know that successful shot-put men tend to be tall and bulky (i.e. heavyweight). Using more than one set of information as a predictive measure involves us in the process called *multiple correlation*. Here we shall only look at the simplest form of multiple correlation, the first order of multiple correlations where we use two sets of information to predict a third. The thing we are trying to predict is called *the criterion*. In the above examples the criterion would be ability to jump high or ability to put the shot. Let us take the example of putting the shot. We have three sets of information about a group of people:

1. How far they can put the shot (the criterion),
2. How tall they are,
3. How heavy they are.

Before we can carry out the process for calculating the multiple R (note it is usually given a capital R), we need to know all of the zero order correlations:

1 with 2 (height with put distance),
1 with 3 (weight with put distance),
2 with 3 (height with weight).

Once we have these three correlations we fit them into the following formula:

$$R_{1.23} = \sqrt{\left[\frac{r_{12}^2 + r_{13}^2 - 2\,(r_{12} \times r_{13} \times r_{23})}{1 - r_{23}^2} \right]}$$

where $R_{1\cdot 23}$ is the multiple correlation between shot-putting, height and weight,

    1 is the criterion: shot-putting ability,

    2 is height,

    3 is weight,

    r_{12} is the r (correlation coefficient) between criterion and height,

    r_{13} is the r between criterion and weight,

    r_{23} is the r between height and weight.

Suppose the values for r_{12}, r_{13}, and r_{23} were as follows:

$$r_{12} = +0.4$$
$$r_{13} = +0.5$$
$$r_{23} = +0.6$$

Then these are put into the formula as shown below:

$$R_{1,23} = \sqrt{\left[\frac{0.4^2 + 0.5^2 - 2(0.4 \times 0.5 \times 0.6)}{1 - 0.6^2}\right]}$$

$$R_{1,23} = \sqrt{\left[\frac{0.16 + 0.25 - 0.24}{1 - 0.36}\right]}$$

$$R_{1,23} = \sqrt{\left[\frac{0.17}{0.64}\right]}$$

$$R_{1,23} = \sqrt{0.2656}$$

$$R_{1,23} = 0.5154$$

This is slightly higher than using either weight or height alone as predictors. Because there is (in this example) a high correlation between height and weight, the multiple is not very much higher than the better of the zero order correlations. Where you have a choice of several zero order predictors you need to bear two things in mind when choosing two for the multiple R.

1. Choose those predictors which have a high correlation with the criterion.

2. Choose predictors which have a low correlation with each other (i.e. they are then telling us something different about the criterion rather than telling us the same thing twice).

In order to achieve both 1 above and 2 above we have to examine the figures. Take the following rs:

A with B $r = + 0.68$
B with C $r = + 0.04$
A with C $r = + 0.94$
A with Z $r = + 0.10$
B with Z $r = + 0.55$
C with Z $r = + 0.32$

We wish to find the best two predictors from A, B and C when predicting Z.

By itself B is the best predictor of Z,
            C is the next best predictor of Z,
            A is the worst predictor of Z.

The correlation between A and C is very high so they have a lot of common ground. The correlation between B and C is very low. Therefore we can expect that the best two predictors of Z in combination will be B and C.

Question 20

The following (hypothetical) correlations were obtained between certain predictors and the likelihood of heart attack. Choose the two most useful for giving a multiple R and calculate what it would be.

(a) r between height and heart attack $= - 0.4$
(b) r between weight and heart attack $= + 0.17$
(c) r between family history of heart trouble and heart attack $= + 0.04$
(d) r between height and weight $= + 0.60$
(e) r between height and family history of heart attack $= - 0.37$
(f) r between weight and family history of heart attack $= + 0.25$

Multiple correlations can be calculated for more than three sets of data. However, this is not usually carried out via the simple zero order correlations and is too complex to be explained here.

Partial Correlation

Another process we may often wish to use is to *hold* constant the effect of one factor whilst seeing what the relationship between the other factors is. For instance, we might want to know how important home background—books available, parental interest in the child's education, parents' actual academic help etc.—is to a child taking the GCE examination, whilst suspecting that IQ also has an important part to play.

The simplest way of offsetting the effect of IQ in a case like this is to divide the children into the two social groups being studied, let us say 'supportive' and 'non-supportive'. Then from each group *match* children of equivalent or very close IQ, discarding the scores of children who cannot be matched.

Supportive Family IQ		Non-supportive Family IQ	
Fred	137	Norma	131
Jean	129	Sid	140
Peter	131	Margaret	129
Joan	124	Bertha	148
Rita	117	Ron	124
Kate	120	Ted	120
Bob	114	Sally	130
Albert	118	Jim	114
Sheila	126	George	116
Claude	115	Lorna	115

Jean and Margaret match	IQ = 129
Peter and Norma match	IQ = 131
Joan and Ron match	IQ = 124
Kate and Ted match	IQ = 120
Bob and Jim match	IQ = 114
Claude and Lorna match	IQ = 115

Fred, Rita, Albert and Sheila in the supportive group, and Sid, Bertha, Sally and George in the non-supportive group have no equivalent partner. Straight away we have cut out 40% of our original sample if we discard these eight people.

Further problems can occur. Suppose we had the following:

Chris	120	Freda	120
Peter	119	Wendy	120
Paul	118	Sandra	118

Which of Freda and Wendy do we discard and which one do we match with Chris? (Chapter 14 gives some of the answers to this problem.) A more satisfactory method is to use the process called *partial correlation*.

In partial correlation no discard is needed and so the full sample is used.

As in multiple correlation the zero-order correlation coefficients are used. Let us use that same example of putting the shot again. This time we want to find out the relationship between ability to put the shot and height *keeping the effect of weight constant*. (Tall men are usually heavier than short men. Is it really their extra height which enables them to put the shot further, or their extra weight, or both?)

We have three sets of information.

1. How far they can put the shot.
2. How tall they are.
3. How heavy they are.

We have calculated the zero-order correlations:

r_{12} = (height with put distance)
r_{13} = (weight with put distance)
r_{23} = (height with weight)

This time the formula we use is:

$$r_{12 \cdot 3} = \frac{r_{12} - r_{13} r_{23}}{\sqrt{[(1 - r_{13}^2)(1 - r_{23}^2)]}}$$

Note that r is not a capital this time and the position of the full stop in $r_{12 \cdot 3}$ allows you to read that as 'the correlation between 1 and 2 holding 3 constant'.

If we use the same values for the zero-order correlation as before, these were:

$$r_{12} = + 0 \cdot 4,$$
$$r_{13} = + 0 \cdot 5,$$
$$r_{23} = + 0 \cdot 6,$$

substituting these in the formula

$$r_{12 \cdot 3} = \frac{0 \cdot 4 - (0 \cdot 5 \times 0 \cdot 6)}{\sqrt{[(1 - 0 \cdot 5^2)(1 - 0 \cdot 6^2)]}}$$

$$= \frac{0 \cdot 4 - 0 \cdot 3}{\sqrt{(0 \cdot 75 \times 0 \cdot 64)}}$$

$$= \frac{0 \cdot 1}{\sqrt{0 \cdot 48}}$$

$$= \frac{0 \cdot 1}{0 \cdot 6928}$$

$$r_{12.3} = 0 \cdot 14$$

In this hypothetical example we can therefore see that the relationship between height and the distance the shot is put is quite small, if we hold the effect of the thrower's weight constant. (Imagine trying to pair off men of equal weight and then getting them to put the shot, as would be required by the alternative matching method.)

Question 21

Given the following (hypothetical) zero-order correlations, calculate the relationship between GCE passes gained and social background for a group of grammar-school boys when the effect of their differing IQs is held constant.

(a) r_{12} between social background and GCE places $= 0 \cdot 50$

(b) r_{13} between social background and IQ $= 0 \cdot 30$

(c) r_{23} between GCE passes and IQ $= 0 \cdot 60$.

CHAPTER ELEVEN

Analysis of variance and the *F*-ratio

As MENTIONED IN Chapter 7, variance can be used to find whether the difference between two small groups is significant or whether the two small groups are really drawn from the same population. However, the big advantage offered by analysis of variance is that it allows us to examine several groups at once to see whether they are significantly different or could have been drawn from the same population. If any one of the subgroups differs significantly from the group as a whole, it will be indicated by the size of the value for *F* which results from the following calculation. However, we should not at this stage know which subgroup was the one significantly different *nor* whether more than one subgroup differed significantly. A *t*-test of significance between each of the various sub groups would then be used to see where the difference lay.

In all tests for significance we are doing the same sort of thing in slightly different ways depending on the size of the samples and the information available. Basically we are taking the two samples and seeing whether the scores in one sample go beyond the *mean* of the other sample. This is called *overlap*. The more one sample overlaps the mean of the other sample the less likely it is that they are significantly different. That is why we tend to compare the gap between the two means and the spread of the scores within the samples. The *F*-ratio employs just such a technique.

(i) Two sets of means are calculated:

1. The two or more subgroup means,
2. The total population mean.

(ii) The deviation of each subgroup mean from the population mean is calculated. These are then squared, multiplied by the number of scores in that subgroup and added together.

The result of (ii) above is then divided by the number of degrees of freedom (i.e. the number of subgroups minus 1). This result is called the *between variance*.

Now we calculate the deviation of each score from its *own* sub-group mean. These are squared and *all* added together. This is then divided by the degrees of freedom (i.e. total number of scores minus the number of subgroups). This result is called the *within variance*.

$$F\text{-ratio} = \frac{\text{between variance}}{\text{within variance}}$$

The *F*-ratio is looked up in Tables 5 and 6.

When the *F*-ratio is equal to or less than 1, then the difference between the subgroups is *not* significant. If it is more than 1 then the difference *may* be significant depending on:

(i) The number of groups being compared,
(ii) The number of scores within each group,
(iii) The degree of confidence desired, e.g. 0·05 or 0·01.

For example, is the difference between the following two sets of marks significant (for simplicity I have only used two subgroups):

Upper Sixth biology mark		Lower Sixth biology mark	
John	37	Rita	48
Peter	41	Jean	45
Freddy	39	Joan	44
Norman	47	Myra	40
		Mavis	43

1. Calculate the three sets of means.

Upper Sixth	Lower Sixth
37	48
41	45
39	44
47	40
———	43
164	———

$$M = \frac{164}{4} = 41, \qquad M = \frac{220}{5} = 44.$$

Total population $= 164 + 220 = 384,$

$$M = \frac{384}{9} = 42 \cdot 67.$$

2. Deviations of the two subgroup means from population mean $= 1 \cdot 67$ and $1 \cdot 33$.

Square these deviations.
$$1 \cdot 67^2 + 1 \cdot 32^2 = 2 \cdot 7889 + 1 \cdot 7689$$
$$= 4 \cdot 5578$$
$$df = \text{number of subgroups} - 1 = 2 - 1 = 1$$
$$\text{Between variance} = \frac{4 \cdot 5578}{1} = 4 \cdot 5578$$

3. Calculate the two sets of deviations from the subgroup means, square them and add them together. Divide by the number of degrees of freedom.

Upper Sixth	Deviation from mean	Deviation²	Lower Sixth	Deviation from mean	Deviation²
37	4	16	48	4	16
41	0	0	45	1	1
39	2	4	44	0	0
47	6	36	40	4	16
			43	1	1
$\sum D^2 = 56$			$\sum D^2 = 34$		

Degrees of freedom $= N - 2 = 7$

$$\text{Within variance} = \frac{56 + 34}{N - 2} = \frac{90}{7} = 12 \cdot 857$$

$$F\text{-ratio} = \frac{4 \cdot 5578}{12 \cdot 857} = 0 \cdot 3545$$

This is less than the *F*-ratio required for significance at the $0 \cdot 05$ level.

Question 22

Do the following GCE results warrant the children continuing in different classes when they enter the sixth form?

5A		5B	
Jane	7 passes	Richard	8 passes
Rita	9 passes	Jean	5 passes
Peter	8 passes	Betty	7 passes
Reggie	6 passes	Philip	8 passes
		Joe	8 passes

5C		5D	
Norma	6 passes	Letty	3 passes
Paul	5 passes	Keith	10 passes
Zena	8 passes	Tilly	6 passes
Ian	7 passes		

Question 23

The following two teams took part in *Top of the Form* representing their two separate schools. Was the difference between them significant?

St. John's School		St. Patrick's School	
Name	Score	Name	Score
Paul	8	Anne	9
Matthew	7	Audrey	12
Mary	11	Sinclair	8
Joan	8	Noel	12

CHAPTER TWELVE
Chi square

CHI-SQUARE DIFFERS FROM the other tests of significance which we have mentioned so far. It does not allow us to compare actual values of scores or measurements in two separate groups. Instead, it allows us to compare the frequencies of scores landing in certain categories. Basically, we compare the number of times something actually happens (the *actual* frequency) with the number of times we expect it to happen (the expected frequency). Chi-square (χ^2), once again, will not tell us why something has happened; it is our job to interpret the results.

There is one draw-back to using Chi-square. It should not be used for calculations where the expected frequency is less than five. For example, if you were on holiday in Rome for the whole of the month of July and decided to compare the number of days on which snow actually fell with the number on which you predicted it would fall, you would probably predict zero days of snow. Because your *predicted frequency* is less than five you cannot use Chi-square.

A technique for correcting this is known as Yates's correction, and is dealt with in more advanced texts.

The formula for Chi-square is very simple:

$$\chi^2 = \sum \left[\frac{(A - E)^2}{E} \right]$$

where A = Actual result (frequency) and E = Expected result (frequency).

Follow this example:

Dave Bedford and Ian Stewart were to have 20 races over various distances from 1500 metres up to 10,000 metres. It was predicted that Bedford would win fifteen and Stewart would win five. In actual practice, Bedford won eleven and Stewart won nine. Was this result significantly different from what was expected?

Make up a table of actual and expected results as follows:

Name	Expected (E) result	Actual (A) result	A—E	(A—E)²	(A—E)² / E
Bedford	15	11	— 4	16	$\frac{16}{15} = 1\cdot07$
Stewart	5	9	+ 4	16	$\frac{16}{5} = 3\cdot2$

$$\chi^2 = \sum \left[\frac{(A - E)^2}{E} \right] = 4\cdot27$$

For this type of use of χ^2 the number of degrees of freedom is one less than the total number of runners. So we look up 4·27 against 1df in Table 4, page 127. For the result to be significantly different from what we expected for one degree of freedom, we find that χ^2 must be greater than 3·84 to be beyond the 0·05 level, and greater than 6·63 to be beyond the 0·01 level. We can therefore interpret our results by saying 'Stewart won significantly more (beyond the 0·05 level) races against Bedford than we had expected'.

Note (1) that degrees of freedom can never be less than one; (2) when expected and actual results are the same, χ^2 will be 0, and there is no point in doing a calculation.

The general rule when calculating degrees of freedom for Chi-square is:

degrees of freedom = (rows — 1) × (columns — 1)

So, if in the previous example there had been four runners, not two, and three types of result (not just a straight winner)—a winner, a tie, and race abandoned—then the degrees of freedom would be the number of runners minus one (i.e. rows — 1) = (4 — 1), multiplied by the number of possible results minus one (i.e. columns — 1) = (3 — 1). Thus the degree of freedom would be 3 × 2 = 6.

If we look at an actual example of χ^2 being used, an article in *Educational Research* of June 1967 serves admirably. G. E. Book-binder investigated the number of children in ESN classes and compares this with the month in which they were born. He reports the following:

Results

Since there is not a uniform birth-rate throughout the year, the figures for live births of Bristol children were obtained for the years 1953–57. From these, expected figures of ESN children for each month of birth have been calculated and compared with the actual numbers found for each month of birth.

TABLE 1: *Months of Birth for Children attending ESN Classes in Ordinary Schools*

	Expected	Actual	E—A
Sept.*	99	80	19
Oct.	104	81	23
Nov.	93	72	21
Dec.	103	91	12
Jan.	116	98	18
Feb.	103	96	7
Mar.	115	118	— 3
Apr.	111	110	1
May	118	124	— 6
June	108	120	— 12
July	112	143	— 31
Aug.*	107	156	— 49

* 1st of September is included in August birthdays because the school year (for age purposes) starts at midnight on September 1st/2nd.

The trend here is unmistakable and is clearly in accord with the expectation of a preponderance of summer-born children in these classes. . . . $\chi^2 = 50 \cdot 8$ (11 df) which means that there is less than one chance in a thousand of such a distribution occurring by chance.

CHAPTER THIRTEEN

Calculation from grouped data

MOST OF THE calculations we have so far considered have assumed we were using discrete and continuous data. That is where each piece of information has been considered as a separate entity. When we have large amounts of data to process it would be very tedious to treat them in that way. The process of grouping data together allows us to consider ten or fifteen groups of information rather than say 100 or 1000 separate pieces of information. Calculation by grouping data has two advantages, speed and ease of handling. It has two disadvantages:

1. Data once in a group is hidden and operates as if at that group midpoint.

2. Information within a group is more likely to be near the population mean than the extreme ends of the range; the process of using the group midpoint as the focus for scores within the group gives the impression (though slight) that scores are spread out more than they actually are.

Bearing these last two points in mind it is probably true to say that where more than 100 cases are being handled the advantages outweigh the disadvantages *unless* you have access to an electric calculator.

Grouping Data
1. The first thing to do with data when grouping is to calculate the *range*.

2. Divide the range into *roughly* fifteen equal *intervals* or *steps*.

3. Find out how many scores lie within each interval.

Given the following data sort it into intervals. (I have used a small number of scores for ease of following the example.)

1	5	7	9	11	8	4	3	17	10	22	8	9	10	14
16	23	18	19	20	24	16	8	3	15	26	28	23		
28	30	12	17	4	25	33	28	37	31	42	45	26		
18	5	43	34	26	37	13	25	33						

1. Calculate the range.

The lowest score is 1.

The highest score is 45.

The range is from 1 to 45 = 45.

2. Divide the range into roughly fifteen equal intervals. (In this case fifteen goes exactly into 45.) Therefore, each interval will be worth three units, e.g. from 1 to 3, 4 to 6, 7 to 9, etc.

3. Find out how many scores lie in each interval. To do this write your intervals in columns and put a tick against the appropriate column, then finally add up the number of ticks in each. You can write your intervals vertically or horizontally, whichever suits you. The number of ticks in each column is called the *frequency* (*f*).

Score	Ticks	f	d	fd	
1 to 3	✓ ✓ ✓	3	− 6	− 18	
4 to 6	✓ ✓ ✓ ✓	4	− 5	− 20	
7 to 9	✓ ✓ ✓ ✓ ✓ ✓	6	− 4	− 24	
10 to 12	✓ ✓ ✓ ✓	4	− 3	− 12	
13 to 15	✓ ✓ ✓	3	− 2	− 6	
16 to 18	✓ ✓ ✓ ✓ ✓ ✓	6	− 1	− 6	
19 to 21	✓ ✓	2	0	0	← Assumed
22 to 24	✓ ✓ ✓ ✓	4	+ 1	+ 4	mean
25 to 27	✓ ✓ ✓ ✓ ✓	5	+ 2	+ 10	interval
28 to 30	✓ ✓ ✓ ✓	4	+ 3	+ 12	
31 to 33	✓ ✓	2	+ 4	+ 8	
34 to 36	✓ ✓	2	+ 5	+ 10	
37 to 39	✓ ✓	2	+ 6	+ 12	
40 to 42	✓	1	+ 7	+ 7	
43 to 45	✓ ✓	2	+ 8	+ 16	

Check that you have the same number of ticks as pieces of original data.

$$3 + 4 + 6 + 4 + 3 + 6 + 2 + 4 + 5 + 4 + 2 + 2 + 2 + 1 + 2 = 50.$$

Calculation of the mean (grouped data)

(Note that here we use the plus and minus signs and *do not* ignore them as in earlier work on standard deviation.)

1. Make a rough guess as to the interval where the mean will be. This is called the *assumed mean interval*.

2. Number each interval going *away* from the assumed mean, which we call zero. Intervals for scores less than the assumed mean interval are negative, those above are positive. Here we are showing how far each interval *deviates* from the assumed mean interval. (See Column marked d in the preceding table.)

Now we use the following formula to calculate the *actual mean.*

$$\text{Actual mean} = \text{assumed mean} + i\frac{(\sum fd)}{N}$$

where i = size of interval,
$\sum$ = 'the sum of',
f = number of scores in the interval,
d = deviation of interval from assumed mean interval,
N = number of cases.

In our example $i = 3$, $N = 50$ and the rest of the information is in the table.

So our formula now looks like this:

$$\text{Actual mean} = \text{assumed mean} + 3\frac{(\sum fd)}{50}$$

$\sum fd$ = the sum of — 18 — 20 — 24 — 12 — 6 — 6
0 4 10 12 8 10 12 7 16
= — 86 + 79
$\sum fd = -7$

$$\text{Actual mean} = \text{assumed mean} + 3\frac{(-7)}{50}$$

$$\text{Actual mean} = \text{assumed mean} - \frac{21}{50}$$

The actual mean is 0·42 below the assumed mean.

What is the assumed mean?

It is the middle of the assumed mean interval. We chose the interval 19 to 21 as our assumed mean interval. The middle of this interval is 20. Therefore our actual mean is 20 — 0·42.

$$\text{Actual mean} = 19·58.$$

Standard Deviation (from grouped data)

The formula for calculating the standard deviation from grouped data is as follows:

$$\sigma = \left\{ \sqrt{\left[\frac{\sum fd^2}{N} - \left(\frac{\sum fd}{N} \right)^2 \right]} \right\} i$$

where
$\sum$ = 'the sum of',

fd = frequency of scores within an interval times the deviation of the interval from the assumed mean interval,

$\dfrac{fd^2}{N}$ = fd times d divided by N,

$\left(\dfrac{\sum fd}{N} \right)^2$ = the sum of all the fds divided by N, then the result squared.

Take the example used for finding the mean. We now need a new row of figures for fd^2. The data then become:

f	d	fd	fd^2	f	d	fd	fd^2
3	− 6	− 18	+ 108	5	+ 2	+ 10	+ 20
4	− 5	− 20	+ 100	4	+ 3	+ 12	+ 36
6	− 4	− 24	+ 96	2	+ 4	+ 8	+ 32
4	− 3	− 12	+ 36	2	+ 5	+ 10	+ 50
3	− 2	− 6	+ 12	2	+ 6	+ 12	+ 72
6	− 1	− 6	+ 6	1	+ 7	+ 7	+ 49
2	0	0	0	2	+ 8	+ 16	+ 128
4	+ 1	+ 4	+ 4				

1. $\dfrac{\sum fd^2}{N} = \dfrac{108 + 100 + 96 + 36 + 12 + 6 + 0}{50} +$

$+ \dfrac{4 + 20 + 36 + 32 + 50 + 72 + 49 + 128}{50}$

$= \dfrac{749}{50}$

2. $\left(\dfrac{\sum fd}{N} \right)^2 = \left(\dfrac{-18 - 20 - 24 - 12 - 6 - 6 - 0}{50} + \right.$

$$+ \frac{4 + 10 + 12 + 8 + 10 + 12 + 7 + 16}{50} \Bigg)^2$$

$$= \left(\frac{-86 + 79}{50} \right)^2$$

$$= \left(\frac{-7}{50} \right)^2$$

3. $i = 3$,

$$\sigma = \left\{ \sqrt{\left[\frac{749}{50} - \left(\frac{-7}{50} \right)^2 \right]} \right\} \times 3$$
$$= [\sqrt{(14 \cdot 98 - 0 \cdot 0196)}] \times 3$$
$$= (\sqrt{14 \cdot 96}) \times 3$$
$$= 3 \cdot 868 \times 3$$
$$= 11 \cdot 6.$$

Question 24

The following marks were gained by the first form at St. Agnes' School in their first year examinations.

217	216	215	214	219	210	227	259	263	200
202	222	232	242	252	231	241	221	211	201
197	284	237	247	205	255	278	246	279	237
191	203	227	241	257	262	293	242	261	217
210	225	238	253	246	213	234	258	272	200

(a) Choose a suitable number of intervals and group the marks accordingly.

(b) Calculate the mean mark for the first form.

(c) Calculate the standard deviation for the first form.

Pearson Product Moment Correlation (from grouped data)

In order to calculate the correlation between two sets of scores we firstly record them in suitable sets of intervals exactly the same as when calculating the mean. The scores are then plotted on a graph with the two assumed mean intervals as the origins. This type of graph is called a *scattergram*.

If scores from both groups fall predominantly in quadrants 1 and 4 the correlation will be negative, if they fall in quadrants 2 and 3 the correlation will be positive.

Let us follow the process through in detail with a very small set of scores. Calculate the correlation between the following marks in art and biology for class 5B at St. Ogg's Grammar School.

Name	Art mark	Biology mark
John	68	70
Fred	65	77
George	61	64
Peter	57	81
Reggie	56	58
Norman	53	85
Philip	48	60
Brian	44	73
Ted	42	51
Sid	40	35
Joe	40	46
Ray	39	42

1. Calculate the range for both sets of scores.

Inclusive range for art $= 39$ to $68 = 30$.
Inclusive range for biology $= 35$ to $85 = 51$.

2. Choose a convenient number of intervals for both sets of scores. (As we have a small number of scores we will only use a small number of intervals.)

Art: 6 intervals worth 5 points each, starting at 39.
Biology: 9 intervals worth 6 points each, starting at 35.

3. Choose your assumed mean intervals.

Art: interval from 54 to 58. Assumed mean $= 56$.
Biology: interval from 59 to 64. Assumed mean $= 61{\cdot}5$.

4. Fill in the data at the left side and bottom of your graph paper *as shown on page* 104.

5. Note that every square on the graph paper has got a number in its corner. This number is the result of multiplying the vertical deviation value by the horizontal deviation value. For example, the top right-hand square is $+ 4$ deviations from the assumed mean interval on the biology scale and $+ 2$ deviations from the assumed mean on the art scale. So, your next task is to calculate this value for each square and put in on the graph.

6. Looking at your original data table you are now able to fill in the ticks in their appropriate squares on the graph. For example, John is 2 deviations above the assumed mean for art, and 1 deviation above the mean for biology. His tick should therefore go in the extreme right-hand column and the fourth row down from the top.

7. Next multiply the value of the square by the number of ticks in it, giving the ringed numbers shown on page 102. Note that the mean intervals have no values in their squares and so the ticks in those two rows are ignored.

8. Add up the total value of the numbers ringed across each row and enter in the separate boxes headed $(f_1 \, d_1)$ $(f_2 \, d_2)$ as shown on page 105.

Biology											
$f_1d_1^2$	f_1d_1	d_1	f_1	i_1	Ticks						
16	4	+4	1	88↑83	√	−12	−8	−4		+4	+8
18	6	+3	2	82↑77	√ √	−9	−6	−3		+3	+6
4	2	+2	1	76↑71	√	−6	−4	−2		+2	+4
1	1	+1	1	70↑65	√	−3	−2	−1		+1	+2
0	0	0	2	64↑59	√ √						
1	−1	−1	1	58↑53	√	+3	+2	+1		−1	−2
4	−2	−2	1	52↑47	√	+6	+4	+2		−2	−4
18	−6	−3	2	46↑41	√ √	+9	+6	+3		−3	−6
16	−4	−4	1	40↑35	√	+12	+8	+4		−4	−8

Assumed mean interval for biology

					Ticks	√ √ √ √	√ √ √	√	√ √	√	√ √
					i_2	39↓43	44↓48	49↓53	54↓58	59↓63	64↓68
					f_2	4	2	1	2	1	2
					d_2	−3	−2	−1	0	+1	+2
					f_2d_2	−12	−4	−1	0	1	4
					$f_2d_2^2$	36	8	1	0	1	8

Art

Assumed mean interval for art

$(f_2d_2)(f_1d_1)$ [ ][ ][ ][ ][ ][ ][ ][ ][ ][ ][ ][ ]

$= \Sigma (f_2d_2)(f_1d_1)$

	Biology	Art
frequency of scores	f_1	f_2
deviation from assumed mean interval	d_1	d_2
size of interval	i_1	i_2

Biology

$f_1d_1^2$	f_1d_1	d_1	f_1	t_1	Ticks						
16	4	4	1	88 ↑ 83	✓	−12	−8	−4 (−4)		+4	+8
18	6	3	2	82 ↑ 77	✓ ✓	−9	−6	−3	✓	+3	+6 (+6)
4	2	2	1	76 ↑ 71	✓	−6	−4 (−4)	−2		+2	+4
1	1	1	1	70 ↑ 65	✓	−3	−2	−1		+1	+2 (+2)
0	0	0	2	64 ↑ 59	✓ ✓		✓			✓	
1	−1	−1	1	58 ↑ 53	✓	+3	+2	+1	✓	−1	−2
4	−2	−2	1	52 ↑ 47	✓	+6 (+6)	+4	+2		−2	−4
18	−6	−3	2	46 ↑ 41	✓ ✓	+9 / (−18)	+6	+3		−3	−6
16	−4	−4	1	40 ↑ 35	✓	+12 / (−12)	+8	+4		−4	−8

	Ticks	✓ ✓ / ✓ ✓	✓ / ✓	✓	✓ / ✓	✓	✓ / ✓
	t_2	39 ↓ 43	44 ↓ 48	49 ↓ 53	54 ↓ 58	59 ↓ 63	64 ↓ 68
	f_2	4	2	1	2	1	2
	d_2	−3	−2	−1	0	+1	+2
	f_2d_2	−12	−4	−1	0	1	4
	$f_2d_2^2$	36	8	1	0	1	8

Art

									total
$(f_1d_1)(f_2d_2)$	−4	+6	−4	+2	0	0	+6	+18	+12	±36

$= \Sigma (f_1d_1)(f_2d_2)$

9.   Calculate $\sum f_1 = \sum f_2 = 12 (=N)$.

Calculate $\sum (f_1 d_1 \checkmark) = + 36 (=$ sum of numbers in rings).

Calculate

$\sum f_1 d_1^2 = 16 + 18 + 4 + 1 + 0 + 1 + 4 + 18 + 16 = 78$

Calculate $\sum f_2 d_2^2 = 36 + 8 + 1 + 0 + 1 + 8 = 54$

Calculate $\dfrac{\sum f_1 d_1}{N} = \dfrac{4+6+2+1+0-1-2-6-4}{12} = \dfrac{0}{12}$

Calculate $\dfrac{\sum f_2 d_2}{N} = \dfrac{-12-4-1+0+1+4}{12} = \dfrac{-12}{12}$

Now we are ready to fit our information into a formula which is:

$$r = \frac{\dfrac{\sum (f_1 d_1 \checkmark)}{N} - \left(\dfrac{\sum f_1 d_1}{N}\right)\left(\dfrac{\sum f_2 d_2}{N}\right)}{\sigma f_1 \times \sigma f_2}$$

$$= \frac{\dfrac{+36}{12} - \left(\dfrac{0}{12}\right) \times \left(\dfrac{-12}{12}\right)}{\sigma f_1 \times \sigma f_2}$$

$$= \frac{3}{\sigma f_1 \times \sigma f_2}$$

10. As we have not previously calculated the standard deviation for either group, we must now do that. Note that in this process we are dealing with deviation scores, so we do not have to multiply the sigmas by the respective values for i.

$$\sigma f_1 = \sqrt{\left[\dfrac{\sum f_1 d_1}{N} - \left(\dfrac{\sum f_1 d_1}{N}\right)^2\right]}$$

$$= \sqrt{\left[\dfrac{78}{12} - \left(\dfrac{0}{12}\right)^2\right]}$$

$$= \sqrt{6 \cdot 5}$$

$$= 2 \cdot 55$$

$$\sigma f_2 = \sqrt{\left[\frac{\Sigma f_2 d_2}{N} - \left(\frac{\Sigma f_2 d_2}{N}\right)^2\right]}$$

$$\sigma f_2 = \sqrt{\left[\frac{54}{12} - \left(\frac{-12}{12}\right)^2\right]}$$
$$= \sqrt{(4 \cdot 5 - 1)}$$
$$= 1 \cdot 87.$$

Substituting the values for σf_1 and σf_2 in the question we get:

$$r = \frac{3}{\sigma f_2 + \sigma f_2}$$
$$= \frac{3}{2 \cdot 55 \times 1 \cdot 87}$$
$$= \frac{3}{4 \cdot 7685}$$
$$= 0 \cdot 629.$$

Clearly it takes a very long time to set up a table like the preceding table. However, once it has been set up, a very large number of items can be quickly dealt with.

Question 25

Calculate the Pearson product moment correlation (grouping method) for the following data.

Student No.	Score on creativity test	Teaching practice mark	Student No.	Score on creativity test	Teaching practice mark
1	62	57	41	57	50
2	40	44	42	58	43
3	56	45	43	52	41
4	65	63	44	73	62
5	67	57	45	63	53
6	70	62	46	64	52
7	56	44	47	77	63
8	66	54	48	75	60
9	65	38	49	80	65
10	76	50	50	67	53
11	82	62	51	66	55
12	67	45	52	76	60
13	59	41	53	74	60
14	58	40	54	66	52
15	71	46	55	66	58
16	68	38	56	83	65
17	75	41	57	92	70
18	93	49	58	48	38
19	64	51	59	56	51
20	86	59	60	67	50
21	65	53	61	67	47
22	64	46	62	116	80
23	54	32	63	77	60
24	49	32	64	58	58
25	63	68	65	72	60
26	63	55	66	56	49
27	72	59	67	56	53
28	56	56	68	49	57
29	56	54	69	67	62
30	67	52	70	43	41
31	70	48	71	76	60
32	57	68	72	74	60
33	55	55	73	48	48
34	63	41	74	32	37
35	55	33	75	66	55
36	50	37	76	83	67
37	68	52	77	48	50
38	82	47	78	67	50
39	78	56	79	72	59
40	50	43	80	49	51

$N = 80$

CHAPTER FOURTEEN
Sampling, sampling errors and statistical misuse

MOST OF THE classroom studies that you do will probably cover the total population and so the errors due to poor sampling or chance will not occur. However, if you are interested in finding out about a larger population the sheer amount of work involved often makes sampling the only practical way.

When we take a sample we assume (within the limits of this book) that the population from which it is drawn is distributed somewhere close to the normal curve for whatever we are measuring. *We also assume* that our sample follows normal distribution. (Indeed we make this assumption for most of our simpler statistical calculation.) There is no way of knowing whether our sample is distributed normally beforehand without knowing already the things we are going to measure, so we do the next best thing and try to get a random sample. With a small population where, say, a 20 per cent sample is to be studied, we can give everyone a number on a raffle ticket, put them in a hat and draw out 20 per cent of the tickets.

With larger populations this would be impossible—for example, a sample census of a small town of 20,000 people. In this case we tend to do such things as open the register of voters (if dealing with adults) stick a pin in the first page and then starting with whichever name the pin goes through select every fifth (for 20 per cent sample tenth for 10 per cent sample, etc.) name throughout the book. It can be argued that only the first name is randomly chosen in this procedure.

Let us look at some of the pitfalls that can occur when selecting a sample group or selecting a control group.

1. Beware of a *pseudo-large* sample, e.g. it is possible to have people in a sample who do not have any effect at all on the thing you are measuring. For instance, suppose your local telephone exchange installs automatic dialling. You wish to see what the public think of the initial inconveniences of the new system. To choose your sample from the electoral roll would mean having

many people in your sample who did not have a telephone at home and therefore probably did not use one regularly or even rarely. These people are 'dead wood', making the sample look larger than it really is; the sample should have been chosen from the telephone directory, or a check question inserted to ensure that only regular users of the new system were included.

2. Beware of sample bias. To give an obvious example, you would not stand at the entrance to the Royal Enclosure at Ascot and ask people going in what their annual income was and then quote the average as being the average income in Britain today.

Less obvious bias comes from assuming that we (that is ourselves, family and friends, workmates, etc.) are average people, with typical attitudes, beliefs and opinions, so instead of using a control group for comparison, we use ourselves.

3. Beware of assuming that a volunteer sample is unbiased, after all something caused them to volunteer whilst others didn't. Always try to have a control group of non-volunteers if this is at all possible.

4. If you are working with a small sample from a large population and get, or hope to get, significance at or above the 0·05 level, it is worthwhile running two samples. For significance to occur at the 0·05 level means that five times out of 100 you could have chosen a non-typical sample. If you get two samples both sig-nificant at the 0·05 level, this likelihood is reduced to $\dfrac{5 \times 5}{100 \times 100}$ or one in 400 times. Alternately why not use a larger sample in the first place?

5. *Misused figures.* Unfortunately so few people understand any statistics that they are apt to take an author's word for *statistical proof.* Where statistics are quoted take a good look at what is being said and apply a bit of ordinary common sense. It may well be that either the figures are wrongly used, or based on criteria which you are unable to accept as valid.

Take the following example from a recent article in a reputable journal which I am sure would never have appeared in its present form if the editor had known any statistics. Two groups of girls (N for each group $= 86$) aged sixteen to twenty were matched for various things. One group had volunteered to go on an Outward Bound Course, the other had not. Approximately nineteen months later the two groups were compared again. The *aim* of the investigation was described as '. . . to present statistical evidence of the influence of Outward Bound Courses on the personality of girls by comparing the differences between initial and final assessment of personality between two groups of girls, one group having attended . . . whilst the other did not attend.' The author then proceeded to analyse the results from Catell's Sixteen Personality Factor Questionnaire. (Remember that at the beginning the groups were matched, on these 16 personality factors amongst other things, so there should have been no statistical difference between them.)

There was *no* statistically significant difference between the two groups nineteen months later, except that the O.B. girls were less conscientious, which the author interprets as meaning they were more unconventional!

At this point the author drops the control group and compares the O.B. girls before and after the nineteen-month gap. They are significantly more stable, etc., after nineteen months and showed a clear improvement in social adaptability. *As there was no significant difference between the two groups this must also be true of the control group.* The author, however, does not bother to say in her summary of results that everything which is true of the O.B. girls is also true of the control girls. The change must be due to a factor or factors *external* to the Outward Bound Course—almost certainly, bearing in mind the ages of the girls, the result of growing more mature with the passage of time. She finishes by quietly picking up the control group again and saying, 'The character-training effects of a [Outward Bound] course are real and can be assessed by comparing their personality development with a control group.'

I find this article particularly saddening because I believe in the Outward Bound Movement and because either through a lack of understanding of the results she found, or a total lack of honesty, the author has set out to *prove* statistically the value of O.B. Courses. Her article is apparently being quoted in all sincerity by others who have not understood the way she uses her evidence.

In short you need to check that the following rules are obeyed when setting up a piece of research or reading a report of some piece of research.

1. Sample should be randomly selected from population.

2. Control should either be randomly selected from the same population or *carefully* matched on the selection criteria.

3. Avoid pseudo-large samples.

4. Beware of sample or control bias.

5. Make sure that the samples being quoted are the ones being compared.

6. Remember that it is just as valid a result to conclude 'No significant difference' or 'No significant improvement', etc., as the reverse—even if it means you have to reject your hypothesis and with it your personal belief in something.

APPENDIX 1
Further problems

Related to Chapter 1

1. What is the mean IQ of the group of schoolboys whose individual scores are given below?

113 127 141 120 110 98 116 107 127 94
86 144 91 111 156 100 109 104 132 103
127 97 120 102 108 93 88 114 104 99
112 104 98 97 131 76 86 95 100

2. What is the median IQ of the group of schoolboys listed above in question 1?

3. In a normal year of 365 days, which day of which month could be described as the median day?

4. What number of days would be described as *modal* when comparing the various months of the year?

Related to Chapter 2

5. Calculate the standard deviation for the following group of IQs:

100 95 86 76 131 97 98 104 112 99
104 114 88 93 108 102 120 97 127 103
132 104 109 100 156 111 91 144 86 94
127 107 116 98 110 120 141 127 113

6. Assuming the above IQs were a random selection measured from a school population of several hundred children, what would you estimate the standard deviation to be for the whole population?

7. Change the following scores to *z*-scores, assuming the mean and standard deviation which you found in question 5.

    100   95   86   111   113   156

Related to Chapter 3

8. Given the following scores for an examination in Latin at St. Joseph's High School, choose a suitable number of steps and plot a histogram. From the shape of the histogram decide whether the scores appear to be fairly normally distributed or skewed. If skewed, how are they skewed?

Name	Score	Name	Score	Name	Score	Name	Score
John	19	Willie	51	Sheila	38	Penny	60
Fred	27	Joe	80	Jane	48	Sue	70
Peter	4	Len	35	Betty	46	Rita	40
Bill	82	George	41	Freda	88	Laura	44
Reg	59	Mike	55	Norma	32	Kate	55
Norman	63	Harry	65	Maggie	53	Celia	65
Tony	47	Cedric	28	Meg	56	Jean	70
Arthur	73	Henry	49	Martha	58	Joan	91
Claude	38	Dick	38	Brenda	61	Daisy	68
Basil	22	Philip	76	Ann	63	Esther	33
Humphrey	67	Mark	63	Janet	57	Gert	41
Ray	61			Jill	16	Hilda	49
Bert	59			Dot	73		

Related to Chapter 5

9. The following table appeared in an article in *Educational Research* of November 1966 where the average scores in a GCE O-level general studies paper were compared for different groups of students who had studied different subjects.

Group	N	Mean score in general studies	Standard deviation
Classics	9	46·9	10·63
Modern language	9	38·2	15·74
English literature	8	50·0	12·07
Arts	8	57·6	9·49

(a) Calculate the standard error of the mean for each of the above groups.

(b) What do the quoted standard deviations tell you about these scores bearing in mind the means and sample sizes?

10. Given a test with a reliability coefficient of $+0.89$ and a standard deviation of 15, calculate the following:

(a) the standard error of measurement for the test

(b) the range of marks within which you would place a boy given an actual test mark of 119 if you were required to be confident at the 0·01 level

(c) the standard error of prediction for the test

(d) What limits would you predict (at the 0·05 level of confidence) for someone who scored 119 on a previous exposure to the test?

Related to Chapter 6

11. In the *British Journal of Educational Psychology* of February 1948 there is an article dealing with selection of pupils for different types of secondary schools. Correlations were calculated between intelligence test results and school records prior to the test. Spearman's rank order formula was used. Part of the table appears at top of page 116.

School	Number of candidates	Coefficient of correlation between IQ test and school record (pre-test placing)
A	44	0·92
B	43	0·77
C	41	0·87
D	40	0·41
E	46	0·87

(a) Was the formula used the correct one for these samples? Justify your conclusion.

(b) Which of the above correlations are significant and at what level?

12. Sixteen boys set out on a twenty-four-hour hike. The time they took to run 100 yards was also measured. The distances they covered and their sprint times are given below. Calculate the Spearman correlation coefficient between the two.

Name	Distance covered in 24 hours (miles)	100 yards sprint time (seconds)
Fred	62	12·0
John	18	11·5
Bill	91	10·8
Harold	38	13·1
Norman	63	12·1
Claude	59	11·9
George	62	12·0
Dick	47	11·3
Albert	49	11·8
Peter	32	12·6
Egbert	56	11·8
Edwin	80	11·2
Matthew	63	11·9
Mark	40	10·9
Tony	93	12·7
Rupert	50	12·1

13. Calculate the Pearson product moment correlation for the following marks given in GCE English literature at O-level and GCE English literature at A-level two years later to the same schoolchildren.

Student number	GCE O-level mark	GCE A-level mark	Student number	GCE O-level mark	GCE A-level mark
1	46	45	29	63	78
2	58	42	30	58	50
3	65	51	31	57	63
4	72	60	32	56	63
5	81	71	33	49	50
6	49	69	34	55	41
7	53	53	35	53	53
8	61	41	36	63	59
9	72	43			
10	77	80			
11	79	68			
12	68	68			
13	51	45			
14	63	60			
15	60	71			
16	33	41			
17	49	44			
18	80	71			
19	37	30			
20	47	41			
21	57	52			
22	62	58			
23	68	66			
24	78	49			
25	74	71			
26	71	71			
27	65	32			
28	61	59			

14. What is the partial tetrachoric correlation between the following scores obtained in the eleven-plus examination for arithmetic and age at which the child learnt to swim?

Child number	11+ arithmetic score	Age at which child could swim years	months
1	68	10	2
2	62	9	0
3	59	6	6
4	62	8	7
5	31	11	0
6	82	9	7
7	32	9	3
8	52	8	1
9	90	8	5
10	11	7	6
11	56	8	4
12	46	4	0
13	66	10	11
14	81	10	1
15	89	9	3
16	70	9	7
17	16	9	6
18	47	10	1
19	37	8	11
20	55	9	4
21	51	10	0
22	59	10	9
23	49	9	8
24	83	11	1
25	27	7	9
26	83	13	0
27	19	9	7
28	42	8	8
29	41	12	1
30	60	11	11

15. When I bought five oranges at shop A for sixpence each they weighed 4 oz, 5 oz, 4 oz, 4½ oz and 5 oz. From shop B five oranges, for sixpence each, weighed 4 oz, 4½ oz, 5 oz, 6 oz and 6½ oz. By weight alone I got better value from shop B; however, was the difference in the weights significant?

16. Two colleges entered teams for an Easter charity walk. Below is given the distance to the nearest kilometre covered by each student. Was the difference between the two teams significant?

College A

Number	Kilometres walked	Number	Kilometres walked
1	·7	24	25·1
2	·9	25	25·2
3	2·8	26	25·3
4	9·6	27	28·0
5	19·3	28	28·2
6	19·4	29	28·9
7	19·4	30	29·0
8	19·4	31	29·7
9	19·5	32	30·0
10	20·0	33	30·1
11	20·0	34	30·1
12	20·1	35	30·9
13	20·2	36	31·1
14	20·8	37	31·1
15	22·6	38	31·8
16	22·7	39	32·6
17	22·8	40	32·7
18	22·9	41	32·7
19	24·1	42	33·0
20	24·2	43	33·8
21	24·3	44	34·0
22	25·0	45	34·7
23	25·0		

College B

Number	Kilometres walked	Number	Kilometres walked
46	·7	81	34·2
47	2·8	82	34·2
48	19·3	83	34·2
49	20·2	84	34·2
50	20·2	85	34·3
51	22·5	86	34·3
52	22·6	87	34·3
53	22·7	88	34·5
54	24·9	89	34·8
55	24·9	90	35·0
56	25·8	91	35·1
57	26·3	92	35·2
58	27·4	93	35·6
59	27·5	94	35·8
60	28·1	95	36·0
61	28·3	96	36·1
62	28·4	97	36·1
63	28·6	98	36·2
64	28·8		
65	29·0		
66	29·2		
67	29·4		
68	29·6		
69	30·0		
70	30·1		
71	30·2		
72	30·3		
73	30·4		
74	30·5		
75	30·6		
76	30·7		
77	31·0		
78	31·0		
79	32·0		
80	33·0		

Related to Chapter 9

17. Convert the following raw scores in the final examinations at Woosnam College of Education to McCall T-scores and decide who should get the over-all academic prize.

Student	Mark for education	Mark for teaching	Mark for main subject Biol.	Eng.	Hist.	Geog.
1	68	B	51			
2	72	B	64			
3	76	C	68			
4	49	C		37		
5	51	D		39		
6	64	A		46		
7	68	B			84	
8	70	C			83	
9	60	C			81	
10	57	D				79
11	48	C				50
12	58	C				28

When changing a letter grade to a numerical value it is usual to make the lowest grade worth 1, the next 2 and so on. Here put $D = 1$, $C = 2$, $B = 3$, $A = 4$.

Related to Chapter 10

18. When a comparison was made between socio-economic class, length of time spent in a particular house and number of children in the family the following hypothetical results were obtained.

Zero order correlation between:

(a) socio-economic class and time spent in the same house
   $r = 0.26$

(b) socio-economic class and size of family
   $r = + 0.01$

(c) time spent in the same house and size of family
   $r = - 0.53$

Calculate the multiple R for these three factors, where time spent in the same house is your criterion.

19. Teams of boys from four schools in Kent took a test of courage and initiative to decide which team should represent the county in an all-England mountain rescue competition. Was the difference between the teams significant? Scores are given below.

School A		School B		School C		School D	
Name	Score	Name	Score	Name	Score	Name	Score
Bob	132	Tom	110	Fred	84	Alan	90
Dick	109	Harry	100	Mike	83	Ted	80
John	101	Albert	90	Colin	71	Pete	65
Ned	92	Sid	59	Doug	60	Stan	53

Related to Chapter 12

20. In an article in *Educational Research* of February 1966, children from different socio-economic family backgrounds were compared to see whether they improved, deteriorated or remained in the same stream that they had been placed in as a result of the eleven-plus examination.

Assuming you predict that the eleven plus had done a good job of selection (i.e. you predict zero change), compare the actual change after five years with your prediction and find whether there is a significant difference between the top three groups and bottom four groups.

Parental occupation category	1	2	3	4	5	6 and 7
Total no. of children in this category—one year	6	9	10	9	41	27
No. of deteriorators during years 1 to 5	0	0	0	2	15	22

Reproduced from *Educational Research*, February 1966, table 2, p.149.

Occupation categories listed above are as follows:

1. Professional and high administrative
2. Managerial and executive
3. Inspectional, supervisory and other non-manual
4. As for 3 but lower grade
5. Skilled manual and routine grades of non-manual

6 and 7. Semi-skilled manual and unskilled manual

APPENDIX 2
Tables

TABLE 1: *Partial Tetrachoric Correlations against Percentage of the Sample in Top Half of Both Tests (N.B. This is quick but only approximate)*

%	r	%	r	%	r	%	r	%	r
45	0·95	37	0·69	29	0·25	21	− 0·25	13	− 0·69
44	0·93	36	0·65	28	0·19	20	− 0·31	12	− 0·73
43	0·91	35	0·60	27	0·13	19	− 0·37	11	− 0·77
42	0·88	34	0·55	26	0·07	18	− 0·43	10	− 0·81
41	0·83	33	0·49	25	0·00	17	− 0·49	9	− 0·83
40	0·81	32	0·43	24	− 0·07	16	− 0·55	8	− 0·88
39	0·77	31	0·37	23	− 0·13	15	− 0·60	7	− 0·91
38	0·73	30	0·31	22	− 0·19	14	− 0·65	6	− 0·93

TABLE 2: *Values of t against Degrees of Freedom for 0·05 Significance and 0·01 Significance Levels*

Degrees of freedom	0·05 significance level	0·01 significance level
1	12·706	63·657
2	4·303	9·925
3	3·182	5·841
4	2·776	4·604
5	2·571	4·032
6	2·447	3·707
7	2·365	3·499
8	2·306	3·355
9	2·262	3·250
10	2·228	3·169
11	2·201	3·106
12	2·179	3·055
13	2·160	3·012
14	2·145	2·977
15	2·131	2·947
16	2·120	2·921
17	2·110	2·898
18	2·101	2·878
19	2·093	2·861
20	2·086	2·845
21	2·080	2·831
22	2·074	2·819
23	2·069	2·807
24	2·064	2·797
25	2·060	2·787
26	2·056	2·779
27	2·052	2·771
28	2·048	2·763
29	2·045	2·756
30	2·042	2·750
∞	1·960	2·576

Reproduced from F. W. Kellaway (ed.), *Penguin–Honeywell Book of Tables*, Penguin, 1968.

TABLE 3A: *Values of Correlation Coefficient Required for Significance at 0·05 and 0·01 Levels for Samples of Various Size (N) [Product Moment Correlations]*

N	5%	1%	N	5%	1%	N	5%	1%	N	5%	1%
10	0·632	0·765	21	0·433	0·549	34	0·339	0·436	65	0·244	0·317
11	0·602	0·735	22	0·423	0·537	36	0·329	0·424	70	0·235	0·306
12	0·576	0·708	23	0·413	0·526	38	0·320	0·413	75	0·227	0·296
13	0·553	0·684	24	0·404	0·515	40	0·312	0·403	80	0·220	0·287
14	0·532	0·661	25	0·396	0·505	42	0·304	0·393	100	0·197	0·256
15	0·514	0·641	26	0·388	0·496	44	0·297	0·384	125	0·176	0·230
16	0·497	0·623	27	0·381	0·487	46	0·291	0·376	150	0·161	0·210
17	0·482	0·606	28	0·374	0·479	48	0·284	0·368	200	0·139	0·182
18	0·468	0·590	29	0·367	0·471	50	0·279	0·361	400	0·098	0·128
19	0·456	0·575	30	0·361	0·463	55	0·265	0·345	1000	0·062	0·081
20	0·444	0·561	32	0·349	0·449	60	0·254	0·330			

Reproduced from E. F. Lindquist, *Statistical Analysis in Educational Research*, Houghton Mifflin, 1940.

TABLE 3B: *Approximate Values of Correlation Coefficient Required for Significance at 0·05 and 0·01 Levels for samples of Various Size (N) [Spearman Rank order correlations]*

N	5%	1%	N	5%	1%	N	5%	1%	N	5%	1%
5	1·00	1·00	9	0·715	0·83	16	0·51	0·665	24	0·415	0·54
6	0·93	0·96	10	0·65	0·795	18	0·48	0·635	26	0·395	0·51
7	0·825	0.92	12	0·615	0·75	20	0·455	0·595	28	0·38	0·49
8	0·78	0·875	14	0·54	0·705	22	0·435	0·565	30	0·36	0·47

TABLE 4: *Distribution of x^2*

Degrees of freedom	0·05	0·01	Degrees of freedom	0·05	0·01
1	3·84	6·63	16	26·30	32·00
2	5·99	9·21	17	27·59	33·41
3	7·81	11·34	18	28·87	34·81
4	9·49	13·28	19	30·14	36·19
5	11·07	15·09	20	31·41	37·57
6	12·59	16·81	21	32·67	38·93
7	14·07	18·48	22	33·92	40·29
8	15·5	20·09	23	35·17	41·64
9	16·9	21·67	24	36·42	42·98
10	18·31	23·21	25	37·65	44·31
11	19·68	24·72	26	38·89	45·64
12	21·03	26·22	27	40·11	46·96
13	22·36	27·69	28	41·34	48·28
14	23·68	29·14	29	42·56	49·59
15	25·00	30·58	30	43·77	50·89

Abridged from F. W. Kellaway (ed.), *Penguin–Honeywell Book of Tables*, Penguin, 1968.

TABLE 5: *Value of* F *Required for* 0·01 *Level of Confidence*

Denominator df = number of scores minus the number of sub-groups	Numerator df = (number of sub-groups − 1)				
	1	2	3	4	5
1	4052·2	4999·5	5403·4	5624·6	5763·6
2	98·50	99·00	99·17	99·25	99·30
3	34·12	30·82	29·46	28·71	28·24
4	21·20	18·00	16·69	15·98	15·52
5	16·26	13·27	12·06	11·39	10·97
6	13·75	10·92	9·78	9·15	8·75
7	12·25	9·55	8·45	7·85	7·46
8	11·26	8·65	7·59	7·01	6·63
9	10·56	8·02	6·99	6·42	6·06
10	10·04	7·56	6·55	5·99	5·64
11	9·65	7·21	6·22	5·67	5·32
12	9·33	6·93	5·95	5·41	5·06
13	9·07	6·70	5·74	5·21	4·86
14	8·86	6·51	5·56	5·04	4·69
15	8·68	6·36	5·42	4·89	4·56
16	8·53	6·23	5·29	4·77	4·44
17	8·40	6·11	5·18	4·67	4·34
18	8·29	6·01	5·09	4·58	4·25
19	8·18	5·93	5·01	4·50	4·17
20	8·10	5·85	4·94	4·43	4·10

Abridged from F. W. Kellaway (ed.), *Penguin–Honeywell Book of Tables*, Penguin, 1968.

TABLE 6: *Value of* F *Required for 0·05 Level of Confidence*

Denominator df = number of scores minus the number of sub-groups	Numerator df = (number of sub-groups − 1)				
	1	2	3	4	5
1	161·45	199·50	215·71	224·58	230·16
2	18·51	19·00	19·16	19·25	19·30
3	10·13	9·55	9·28	9·12	9·01
4	7·71	6·94	6·59	6·39	6·26
5	6·61	5·79	5·41	5·19	5·05
6	5·99	5·14	4·76	4·53	4·39
7	5·59	4·74	4·35	4·12	3·97
8	5·32	4·46	4·07	3·84	3·69
9	5·12	4·26	3·86	3·63	3·48
10	4·96	4·10	3·71	3·48	3·33
11	4·84	3·98	3·59	3·36	3·20
12	4·75	3·89	3·49	3·26	3·11
13	4·67	3·81	3·41	3·18	3·03
14	4·60	3·74	3·34	3·11	2·96
15	4·54	3·68	3·29	3·06	2·90
16	4·49	3·63	3·24	3·01	2·85
17	4·45	3·59	3·20	2·96	2·81
18	4·41	3·55	3·16	2·93	2·77
19	4·38	3·52	3·13	2·90	2·74
20	4·35	3·49	3·10	2·87	2·71

Abridged from F. W. Kellaway (ed.), *Penguin–Honeywell Book of Tables*, Penguin, 1968.

TABLE 7: *Square Roots. From 1 to 10*

	0	1	2	3	4	5	6	7	8	9	Mean differences 1 2 3	4 5 6	7 8 9
1·0	1·000	1·005	1·010	1·015	1·020	1·025	1·030	1·034	1·039	1·044	0 1 1	2 2 3	3 4 4
1·1	1·049	1·054	1·058	1·063	1·068	1·072	1·077	1·082	1·086	1·091	0 1 1	2 2 3	3 4 4
1·2	1·095	1·100	1·105	1·109	1·114	1·118	1·122	1·127	1·131	1·136	0 1 1	2 2 3	3 4 4
1·3	1·140	1·145	1·149	1·153	1·158	1·162	1·166	1·170	1·175	1·179	0 1 1	2 2 3	3 3 4
1·4	1·183	1·187	1·192	1·196	1·200	1·204	1·208	1·212	1·217	1·221	0 1 1	2 2 2	3 3 4
1·5	1·225	1·229	1·233	1·237	1·241	1·245	1·249	1·253	1·257	1·261	0 1 1	2 2 2	3 3 4
1·6	1·265	1·269	1·273	1·277	1·281	1·285	1·288	1·292	1·296	1·300	0 1 1	2 2 2	3 3 3
1·7	1·304	1·308	1·311	1·315	1·319	1·323	1·327	1·330	1·334	1·338	0 1 1	2 2 2	3 3 3
1·8	1·342	1·345	1·349	1·353	1·356	1·360	1·364	1·367	1·371	1·375	0 1 1	1 2 2	3 3 3
1·9	1·378	1·382	1·386	1·389	1·393	1·396	1·400	1·404	1·407	1·411	0 1 1	1 2 2	3 3 3
2·0	1·414	1·418	1·421	1·425	1·428	1·432	1·435	1·439	1·442	1·446	0 1 1	1 2 2	2 3 3
2·1	1·449	1·453	1·456	1·459	1·463	1·466	1·470	1·473	1·476	1·480	0 1 1	1 2 2	2 3 3
2·2	1·483	1·487	1·490	1·493	1·497	1·500	1·503	1·507	1·510	1·513	0 1 1	1 2 2	2 3 3
2·3	1·517	1·520	1·523	1·526	1·530	1·533	1·536	1·539	1·543	1·546	0 1 1	1 2 2	2 3 3
2·4	1·549	1·552	1·556	1·559	1·562	1·565	1·568	1·572	1·575	1·578	0 1 1	1 2 2	2 3 3
2·5	1·581	1·584	1·587	1·591	1·594	1·597	1·600	1·603	1·606	1·609	0 1 1	1 2 2	2 3 3
2·6	1·612	1·616	1·619	1·622	1·625	1·628	1·631	1·634	1·637	1·640	0 1 1	1 2 2	2 2 3
2·7	1·643	1·646	1·649	1·652	1·655	1·658	1·661	1·664	1·667	1·670	0 1 1	1 2 2	2 2 3
2·8	1·673	1·676	1·679	1·682	1·685	1·688	1·691	1·694	1·697	1·700	0 1 1	1 1 2	2 2 3
2·9	1·703	1·706	1·709	1·712	1·715	1·718	1·720	1·723	1·726	1·729	0 1 1	1 1 2	2 2 3
3·0	1·732	1·735	1·738	1·741	1·744	1·746	1·749	1·752	1·755	1·758	0 1 1	1 1 2	2 2 3
3·1	1·761	1·764	1·766	1·769	1·772	1·775	1·778	1·780	1·783	1·786	0 1 1	1 1 2	2 2 3
3·2	1·789	1·792	1·794	1·797	1·800	1·803	1·806	1·808	1·811	1·814	0 1 1	1 1 2	2 2 2
3·3	1·817	1·819	1·822	1·825	1·828	1·830	1·833	1·836	1·838	1·841	0 1 1	1 1 2	2 2 2
3·4	1·844	1·847	1·849	1·852	1·855	1·857	1·860	1·863	1·865	1·868	0 1 1	1 1 2	2 2 2
3·5	1·871	1·873	1·876	1·879	1·881	1·884	1·887	1·889	1·892	1·895	0 1 1	1 1 2	2 2 2
3·6	1·897	1·900	1·903	1·905	1·908	1·910	1·913	1·916	1·918	1·921	0 1 1	1 1 2	2 2 2
3·7	1·924	1·926	1·929	1·931	1·934	1·936	1·939	1·942	1·944	1·947	0 1 1	1 1 2	2 2 2
3·8	1·949	1·952	1·954	1·957	1·960	1·962	1·965	1·967	1·970	1·972	0 1 1	1 1 2	2 2 2
3·9	1·975	1·977	1·980	1·982	1·985	1·987	1·990	1·992	1·995	1·997	0 1 1	1 1 2	2 2 2
4·0	2·000	2·002	2·005	2·007	2·010	2·012	2·015	2·017	2·020	2·022	0 0 1	1 1 1	2 2 2
4·1	2·025	2·027	2·030	2·032	2·035	2·037	2·040	2·042	2·045	2·047	0 0 1	1 1 1	2 2 2
4·2	2·049	2·052	2·054	2·057	2·059	2·062	2·064	2·066	2·069	2·071	0 0 1	1 1 1	2 2 2
4·3	2·074	2·076	2·078	2·081	2·083	2·086	2·088	2·090	2·093	2·095	0 0 1	1 1 1	2 2 2
4·4	2·098	2·100	2·102	2·105	2·107	2·110	2·112	2·114	2·117	2·119	0 0 1	1 1 1	2 2 2
4·5	2·121	2·124	2·126	2·128	2·131	2·133	2·135	2·138	2·140	2·142	0 0 1	1 1 1	2 2 2
4·6	2·145	2·147	2·149	2·152	2·154	2·156	2·159	2·161	2·163	2·166	0 0 1	1 1 1	2 2 2
4·7	2·168	2·170	2·173	2·175	2·177	2·179	2·182	2·184	2·186	2·189	0 0 1	1 1 1	2 2 2
4·8	2·191	2·193	2·195	2·198	2·200	2·202	2·205	2·207	2·209	2·211	0 0 1	1 1 1	2 2 2
4·9	2·214	2·216	2·218	2·220	2·223	2·225	2·227	2·229	2·232	2·234	0 0 1	1 1 1	2 2 2
5·0	2·236	2·238	2·241	2·243	2·245	2·247	2·249	2·252	2·254	2·256	0 0 1	1 1 1	2 2 2
5·1	2·258	2·261	2·263	2·265	2·267	2·269	2·272	2·274	2·276	2·278	0 0 1	1 1 1	2 2 2
5·2	2·280	2·283	2·285	2·287	2·289	2·291	2·293	2·296	2·298	2·300	0 0 1	1 1 1	2 2 2
5·3	2·302	2·304	2·307	2·309	2·311	2·313	2·315	2·317	2·319	2·322	0 0 1	1 1 1	2 2 2
5·4	2·324	2·326	2·328	2·330	2·332	2·335	2·337	2·339	2·341	2·343	0 0 1	1 1 1	1 2 2

TABLE 7: (*cont.*)

	0	1	2	3	4	5	6	7	8	9	Mean differences 1 2 3	4 5 6	7 8 9
5·5	2·345	2·347	2·349	2·352	2·354	2·356	2·358	2·360	2·362	2·364	0 0 1	1 1 1	1 2 2
5·6	2·366	2·369	2·371	2·373	2·375	2·377	2·379	2·381	2·383	2·385	0 0 1	1 1 1	1 2 2
5·7	2·387	2·390	2·392	2·394	2·396	2·398	2·400	2·402	2·404	2·406	0 0 1	1 1 1	1 2 2
5·8	2·408	2·410	2·412	2·415	2·417	2·419	2·421	2·423	2·425	2·427	0 0 1	1 1 1	1 2 2
5·9	2·429	2·431	2·433	2·435	2·437	2·439	2·441	2·443	2·445	2·447	0 0 1	1 1 1	1 2 2
6·0	2·449	2·452	2·454	2·456	2·458	2·460	2·462	2·464	2·466	2·468	0 0 1	1 1 1	1 2 2
6·1	2·470	2·472	2·474	2·476	2·478	2·480	2·482	2·484	2·486	2·488	0 0 1	1 1 1	1 2 2
6·2	2·490	2·492	2·494	2·496	2·498	2·500	2·502	2·504	2·506	2·508	0 0 1	1 1 1	1 2 2
6·3	2·510	2·512	2·514	2·516	2·518	2·520	2·522	2·524	2·526	2·528	0 0 1	1 1 1	1 2 2
6·4	2·530	2·532	2·534	2·536	2·538	2·540	2·542	2·544	2·546	2·548	0 0 1	1 1 1	1 2 2
6·5	2·550	2·551	2·553	2·555	2·557	2·559	2·561	2·563	2·565	2·567	0 0 1	1 1 1	1 2 2
6·6	2·569	2·571	2·573	2·575	2·577	2·579	2·581	2·583	2·585	2·587	0 0 1	1 1 1	1 2 2
6·7	2·588	2·590	2·592	2·594	2·596	2·598	2·600	2·602	2·604	2·606	0 0 1	1 1 1	1 2 2
6·8	2·608	2·610	2·612	2·613	2·615	2·617	2·619	2·621	2·623	2·625	0 0 1	1 1 1	1 2 2
6·9	2·627	2·629	2·631	2·632	2·634	2·636	2·638	2·640	2·642	2·644	0 0 1	1 1 1	1 2 2
7·0	2·646	2·648	2·650	2·651	2·653	2·655	2·657	2·659	2·661	2·663	0 0 1	1 1 1	1 2 2
7·1	2·665	2·666	2·668	2·670	2·672	2·674	2·676	2·678	2·680	2·681	0 0 1	1 1 1	1 1 2
7·2	2·683	2·685	2·687	2·689	2·691	2·693	2·694	2·696	2·698	2·700	0 0 1	1 1 1	1 1 2
7·3	2·702	2·704	2·706	2·707	2·709	2·711	2·713	2·715	2·717	2·718	0 0 1	1 1 1	1 1 2
7·4	2·720	2·722	2·724	2·726	2·728	2·729	2·731	2·733	2·735	2·737	0 0 1	1 1 1	1 1 2
7·5	2·739	2·740	2·742	2·744	2·746	2·748	2·750	2·751	2·753	2·755	0 0 1	1 1 1	1 1 2
7·6	2·757	2·759	2·760	2·762	2·764	2·766	2·768	2·769	2·771	2·773	0 0 1	1 1 1	1 1 2
7·7	2·775	2·777	2·778	2·780	2·782	2·784	2·786	2·787	2·789	2·791	0 0 1	1 1 1	1 1 2
7·8	2·793	2·795	2·796	2·798	2·800	2·802	2·804	2·805	2·807	2·809	0 0 1	1 1 1	1 1 2
7·9	2·811	2·812	2·814	2·816	2·818	2·820	2·821	2·823	2·825	2·827	0 0 1	1 1 1	1 1 2
8·0	2·828	2·830	2·832	2·834	2·835	2·837	2·839	2·841	2·843	2·844	0 0 1	1 1 1	1 1 2
8·1	2·846	2·848	2·850	2·851	2·853	2·855	2·857	2·858	2·860	2·862	0 0 1	1 1 1	1 1 2
8·2	2·864	2·865	2·867	2·869	2·871	2·872	2·874	2·876	2·877	2·879	0 0 1	1 1 1	1 1 2
8·3	2·881	2·883	2·884	2·886	2·888	2·890	2·891	2·893	2·895	2·897	0 0 1	1 1 1	1 1 2
8·4	2·898	2·900	2·902	2·903	2·905	2·907	2·909	2·910	2·912	2·914	0 0 1	1 1 1	1 1 2
8·5	2·915	2·917	2·919	2·921	2·922	2·924	2·926	2·927	2·929	2·931	0 0 1	1 1 1	1 1 2
8·6	2·933	2·934	2·936	2·938	2·939	2·941	2·943	2·944	2·946	2·948	0 0 1	1 1 1	1 1 2
8·7	2·950	2·951	2·953	2·955	2·956	2·958	2·960	2·961	2·963	2·965	0 0 1	1 1 1	1 1 2
8·8	2·966	2·968	2·970	2·972	2·973	2·975	2·977	2·978	2·980	2·982	0 0 1	1 1 1	1 1 2
8·9	2·983	2·985	2·987	2·988	2·990	2·992	2·993	2·995	2·997	2·998	0 0 1	1 1 1	1 1 2
9·0	3·000	3·002	3·003	3·005	3·007	3·008	3·010	3·012	3·013	3·015	0 0 0	1 1 1	1 1 1
9·1	3·017	3·018	3·020	3·022	3·023	3·025	3·027	3·028	3·030	3·032	0 0 0	1 1 1	1 1 1
9·2	3·033	3·035	3·036	3·038	3·040	3·041	3·043	3·045	3·046	3·048	0 0 0	1 1 1	1 1 1
9·3	3·050	3·051	3·053	3·055	3·056	3·058	3·059	3·061	3·063	3·064	0 0 0	1 1 1	1 1 1
9·4	3·066	3·068	3·069	3·071	3·072	3·074	3·076	3·077	3·079	3·081	0 0 0	1 1 1	1 1 1
9·5	3·082	3·084	3·085	3·087	3·089	3·090	3·092	3·094	3·095	3·097	0 0 0	1 1 1	1 1 1
9·6	3·098	3·100	3·102	3·103	3·105	3·106	3·108	3·110	3·111	3·113	0 0 0	1 1 1	1 1 1
9·7	3·114	3·116	3·118	3·119	3·121	3·122	3·124	3·126	3·127	3·129	0 0 0	1 1 1	1 1 1
9·8	3·130	3·132	3·134	3·135	3·137	3·138	3·140	3·142	3·143	3·145	0 0 0	1 1 1	1 1 1
9·9	3·146	3·148	3·150	3·151	3·153	3·154	3·156	3·158	3·159	3·161	0 0 0	1 1 1	1 1 1

TABLE 8: *Square Roots. From 10 to 100*

	0	1	2	3	4	5	6	7	8	9	Mean differences 1 2 3	4 5 6	7 8 9
10	3·162	3·178	3·194	3·209	3·225	3·240	3·256	3·271	3·286	3·302	2 3 5	6 8 9	11 12 14
11	3·317	3·332	3·347	3·362	3·376	3·391	3·406	3·421	3·435	3·450	1 3 4	6 7 9	10 12 13
12	3·464	3·479	3·493	3·507	3·521	3·536	3·550	3·564	3·578	3·592	1 3 4	6 7 8	10 11 13
13	3·606	3·619	3·633	3·647	3·661	3·674	3·688	3·701	3·715	3·728	1 3 4	5 7 8	10 11 12
14	3·742	3·755	3·768	3·782	3·795	3·808	3·821	3·834	3·847	3·860	1 3 4	5 7 8	9 11 12
15	3·873	3·886	3·899	3·912	3·924	3·937	3·950	3·962	3·975	3·987	1 3 4	5 6 8	9 10 11
16	4·000	4·012	4·025	4·037	4·050	4·062	4·074	4·087	4·099	4·111	1 2 4	5 6 7	9 10 11
17	4·123	4·135	4·147	4·159	4·171	4·183	4·195	4·207	4·219	4·231	1 2 4	5 6 7	8 10 11
18	4·243	4·254	4·266	4·278	4·290	4·301	4·313	4·324	4·336	4·347	1 2 3	5 6 7	8 9 10
19	4·359	4·370	4·382	4·393	4·405	4·416	4·427	4·438	4·450	4·461	1 2 3	5 6 7	8 9 10
20	4·472	4·483	4·494	4·506	4·517	4·528	4·539	4·550	4·561	4·572	1 2 3	4 6 7	8 9 10
21	4·583	4·593	4·604	4·615	4·626	4·637	4·648	4·658	4·669	4·680	1 2 3	4 5 6	8 9 10
22	4·690	4·701	4·712	4·722	4·733	4·743	4·754	4·764	4·775	4·785	1 2 3	4 5 6	7 8 9
23	4·796	4·806	4·817	4·827	4·837	4·848	4·858	4·868	4·879	4·889	1 2 3	4 5 6	7 8 9
24	4·899	4·909	4·919	4·930	4·940	4·950	4·960	4·970	4·980	4·990	1 2 3	4 5 6	7 8 9
25	5·000	5·010	5·020	5·030	5·040	5·050	5·060	5·070	5·079	5·089	1 2 3	4 5 6	7 8 9
26	5·099	5·109	5·119	5·128	5·138	5·148	5·158	5·167	5·177	5·187	1 2 3	4 5 6	7 8 9
27	5·196	5·206	5·215	5·225	5·235	5·244	5·254	5·263	5·273	5·282	1 2 3	4 5 6	7 8 9
28	5·292	5·301	5·310	5·320	5·329	5·339	5·348	5·357	5·367	5·376	1 2 3	4 5 6	7 7 8
29	5·385	5·394	5·404	5·413	5·422	5·431	5·441	5·450	5·459	5·468	1 2 3	4 5 5	6 7 8
30	5·477	5·486	5·495	5·505	5·514	5·523	5·532	5·541	5·550	5·559	1 2 3	4 4 5	6 7 8
31	5·568	5·577	5·586	5·595	5·604	5·612	5·621	5·630	5·639	5·648	1 2 3	3 4 5	6 7 8
32	5·657	5·666	5·675	5·683	5·692	5·701	5·710	5·718	5·727	5·736	1 2 3	3 4 5	6 7 8
33	5·745	5·753	5·762	5·771	5·779	5·788	5·797	5·805	5·814	5·822	1 2 3	3 4 5	6 7 8
34	5·831	5·840	5·848	5·857	5·865	5·874	5·882	5·891	5·899	5·908	1 2 3	3 4 5	6 7 8
35	5·916	5·925	5·933	5·941	5·950	5·958	5·967	5·975	5·983	5·992	1 2 2	3 4 5	6 7 8
36	6·000	6·008	6·017	6·025	6·033	6·042	6·050	6·058	6·066	6·075	1 2 2	3 4 5	6 7 7
37	6·083	6·091	6·099	6·107	6·116	6·124	6·132	6·140	6·148	6·156	1 2 2	3 4 5	6 7 7
38	6·164	6·173	6·181	6·189	6·197	6·205	6·213	6·221	6·229	6·237	1 2 2	3 4 5	6 6 7
39	6·245	6·253	6·261	6·269	6·277	6·285	6·293	6·301	6·309	6·317	1 2 2	3 4 5	6 6 7
40	6·325	6·332	6·340	6·348	6·356	6·364	6·372	6·380	6·387	6·395	1 2 2	3 4 5	6 6 7
41	6·403	6·411	6·419	6·427	6·434	6·442	6·450	6·458	6·465	6·473	1 2 2	3 4 5	5 6 7
42	6·481	6·488	6·496	6·504	6·512	6·519	6·527	6·535	6·542	6·550	1 2 2	3 4 5	5 6 7
43	6·557	6·565	6·573	6·580	6·588	6·595	6·603	6·611	6·618	6·626	1 2 2	3 4 5	5 6 7
44	6·633	6·641	6·648	6·656	6·663	6·671	6·678	6·686	6·693	6·701	1 2 2	3 4 5	5 6 7
45	6·708	6·716	6·723	6·731	6·738	6·745	6·753	6·760	6·768	6·775	1 1 2	3 4 4	5 6 7
46	6·782	6·790	6·797	6·804	6·812	6·819	6·826	6·834	6·841	6·848	1 1 2	3 4 4	5 6 7
47	6·856	6·863	6·870	6·877	6·885	6·892	6·899	6·907	6·914	6·921	1 1 2	3 4 4	5 6 7
48	6·928	6·935	6·943	6·950	6·957	6·964	6·971	6·979	6·986	6·993	1 1 2	3 4 4	5 6 6
49	7·000	7·007	7·014	7·021	7·029	7·036	7·043	7·050	7·057	7·064	1 1 2	3 4 4	5 6 6
50	7·071	7·078	7·085	7·092	7·099	7·106	7·113	7·120	7·127	7·134	1 1 2	3 4 4	5 6 6
51	7·141	7·148	7·155	7·162	7·169	7·176	7·183	7·190	7·197	7·204	1 1 2	3 4 4	5 6 6
52	7·211	7·218	7·225	7·232	7·239	7·246	7·253	7·259	7·266	7·273	1 1 2	3 3 4	5 6 6
53	7·280	7·287	7·294	7·301	7·308	7·314	7·321	7·328	7·335	7·342	1 1 2	3 3 4	5 5 6
54	7·348	7·355	7·362	7·369	7·376	7·382	7·389	7·396	7·403	7·409	1 1 2	3 3 4	5 5 6

TABLE 8: (cont.)

	0	1	2	3	4	5	6	7	8	9	1	2	3	4	5	6	7	8	9
											— Mean differences								
55	7·416	7·423	7·430	7·436	7·443	7·450	7·457	7·463	7·470	7·477	1	1	2	3	3	4	5	5	6
56	7·483	7·490	7·497	7·503	7·510	7·517	7·523	7·530	7·537	7·543	1	1	2	3	3	4	5	5	6
57	7·550	7·556	7·563	7·570	7·576	7·583	7·589	7·596	7·603	7·609	1	1	2	3	3	4	5	5	6
58	7·616	7·622	7·629	7·635	7·642	7·649	7·655	7·662	7·668	7·675	1	1	2	3	3	4	5	5	6
59	7·681	7·688	7·694	7·701	7·707	7·714	7·720	7·727	7·733	7·740	1	1	2	3	3	4	4	5	6
60	7·746	7·752	7·759	7·765	7·772	7·778	7·785	7·791	7·797	7·804	1	1	2	3	3	4	4	5	6
61	7·810	7·817	7·823	7·829	7·836	7·842	7·849	7·855	7·861	7·868	1	1	2	3	3	4	4	5	6
62	7·874	7·880	7·887	7·893	7·899	·7·906	7·912	7·918	7·925	7·931	1	1	2	3	3	4	4	5	6
63	7·937	7·944	7·950	7·956	7·962	7·969	7·975	7·981	7·987	7·994	1	1	2	3	3	4	4	5	6
64	8·000	8·006	8·012	8·019	8·025	8·031	8·037	8·044	8·050	8·056	1	1	2	2	3	4	4	5	6
65	8·062	8·068	8·075	8·081	8·087	8·093	8·099	8·106	8·112	8·118	1	1	2	2	3	4	4	5	6
66	8·124	8·130	8·136	8·142	8·149	8·155	8·161	8·167	8·173	8·179	1	1	2	2	3	4	4	5	5
67	8·185	8·191	8·198	8·204	8·210	8·216	8·222	8·228	8·234	8·240	1	1	2	2	3	4	4	5	5
68	8·246	8·252	8·258	8·264	8·270	8·276	8·283	8·289	8·295	8·301	1	1	2	2	3	4	4	5	5
69	8·307	8·313	8·319	8·325	8·331	8·337	8·343	8·349	8·355	8·361	1	1	2	2	3	4	4	5	5
70	8·367	8·373	8·379	8·385	8·390	8·396	8·402	8·408	8·414	8·420	1	1	2	2	3	4	4	5	5
71	8·426	8·432	8·438	8·444	8·450	8·456	8·462	8·468	8·473	8·479	1	1	2	2	3	4	4	5	5
72	8·485	8·491	8·497	8·503	8·509	8·515	8·521	8·526	8·532	8·538	1	1	2	2	3	3	4	5	5
73	8·544	8·550	8·556	8·562	8·567	8·573	8·579	8·585	8·591	8·597	1	1	2	2	3	3	4	5	5
74	8·602	8·608	8·614	8·620	8·626	8·631	8·637	8·643	8·649	8·654	1	1	2	2	3	3	4	5	5
75	8·660	8·666	8·672	8·678	8·683	8·689	8·695	8·701	8·706	8·712	1	1	2	2	3	3	4	5	5
76	8·718	8·724	8·729	8·735	8·741	8·746	8·752	8·758	8·764	8·769	1	1	2	2	3	3	4	5	5
77	8·775	8·781	8·786	8·792	8·798	8·803	8·809	8·815	8·820	8·826	1	1	2	2	3	3	4	4	5
78	8·832	8·837	8·843	8·849	8·854	8·860	8·866	8·871	8·877	8·883	1	1	2	2	3	3	4	4	5
79	8·888	8·894	8·899	8·905	8·911	8·916	8·922	8·927	8·933	8·939	1	1	2	2	3	3	4	4	5
80	8·944	8·950	8·955	8·961	8·967	8·972	8·978	8·983	8·989	8·994	1	1	2	2	3	3	4	4	5
81	9·000	9·006	9·011	9·017	9·022	9·028	9·033	9·039	9·044	9·050	1	1	2	2	3	3	4	4	5
82	9·055	9·061	9·066	9·072	9·077	9·083	9·088	9·094	9·099	9·105	1	1	2	2	3	3	4	4	5
83	9·110	9·116	9·121	9·127	9·132	9·138	9·143	9·149	9·154	9·160	1	1	2	2	3	3	4	4	5
84	9·165	9·171	9·176	9·182	9·187	9·192	9·198	9·203	9·209	9·214	1	1	2	2	3	3	4	4	5
85	9·220	9·225	9·230	9·236	9·241	9·247	9·252	9·257	9·263	9·268	1	1	2	2	3	3	4	4	5
86	9·274	9·279	9·284	9·290	9·295	9·301	9·306	9·311	9·317	9·322	1	1	2	2	3	3	4	4	5
87	9·327	9·333	9·338	9·343	9·349	9·354	9·359	9·365	9·370	9·375	1	1	2	2	3	3	4	4	5
88	9·381	9·386	9·391	9·397	9·402	9·407	9·413	9·418	9·423	9·429	1	1	2	2	3	3	4	4	5
89	9·434	9·439	9·445	9·450	9·455	9·460	9·466	9·471	9·476	9·482	1	1	2	2	3	3	4	4	5
90	9·487	9·492	9·497	9·503	9·508	9·513	9·518	9·524	9·529	9·534	1	1	2	2	3	3	4	4	5
91	9·539	9·545	9·550	9·555	9·560	9·566	9·571	9·576	9·581	9·586	1	1	2	2	3	3	4	4	5
92	9·592	9·597	9·602	9·607	9·612	9·618	9·623	9·628	9·633	9·638	1	1	2	2	3	3	4	4	5
93	9·644	9·649	9·654	9·659	9·664	9·670	9·675	9·680	9·685	9·690	1	1	2	2	3	3	4	4	5
94	9·695	9·701	9·706	9·711	9·716	9·721	9·726	9·731	9·737	9·742	1	1	2	2	3	3	4	4	5
95	9·747	9·752	9·757	9·762	9·767	9·772	9·778	9·783	9·788	9·793	1	1	2	2	3	3	4	4	5
96	9·798	9·803	9·808	9·813	9·818	9·823	9·829	9·834	9·839	9·844	1	1	2	2	3	3	4	4	5
97	9·849	9·854	9·859	9·864	9·869	9·874	9·879	9·884	9·889	9·894	1	1	2	2	3	3	4	4	5
98	9·899	9·905	9·910	9·915	9·920	9·925	9·930	9·935	9·940	9·945	0	1	1	2	2	3	3	4	4
99	9·950	9·955	9·960	9·965	9·970	9·975	9·980	9·985	9·990	9·995	0	1	1	2	2	3	3	4	4

TABLE 9: *Squares of Numbers*

	0	1	2	3	4	5	6	7	8	9	Mean differences								
											1	2	3	4	5	6	7	8	9
1·0	1·000	1·020	1·040	1·061	1·082	1·103	1·124	1·145	1·166	1·188	2	4	6	8	10	13	15	17	19
1·1	1·210	1·232	1·254	1·277	1·300	1·323	1·346	1·369	1·392	1·416	2	5	7	9	11	14	16	18	21
1·2	1·440	1·464	1·488	1·513	1·538	1·563	1·588	1·613	1·638	1·664	2	5	7	10	12	15	17	20	22
1·3	1·690	1·716	1·742	1·769	1·796	1·823	1·850	1·877	1·904	1·932	3	5	8	11	13	16	19	22	24
1·4	1·960	1·988	2·016	2·045	2·074	2·103	2·132	2·161	2·190	2·220	3	6	9	12	14	17	20	23	26
1·5	2·250	2·280	2·310	2·341	2·372	2·403	2·434	2·465	2·496	2·528	3	6	9	12	15	19	22	25	28
1·6	2·560	2·592	2·624	2·657	2·690	2·723	2·756	2·789	2·822	2·856	3	7	10	13	16	20	23	26	30
1·7	2·890	2·924	2·958	2·993	3·028	3·063	3·098	3·133	3·168	3·204	3	7	10	14	17	21	24	28	31
1·8	3·240	3·276	3·312	3·349	3·386	3·423	3·460	3·497	3·534	3·572	4	7	11	15	18	22	26	30	33
1·9	3·610	3·648	3·686	3·725	3·764	3·803	3·842	3·881	3·920	3·960	4	8	12	16	19	23	27	31	35
2·0	4·000	4·040	4·080	4·121	4·162	4·203	4·244	4·285	4·326	4·368	4	8	12	16	20	25	29	33	37
2·1	4·410	4·452	4·494	4·537	4·580	4·623	4·666	4·709	4·752	4·796	4	9	13	17	21	26	30	34	39
2·2	4·840	4·884	4·928	4·973	5·018	5·063	5·108	5·153	5·198	5·244	4	9	13	18	22	27	31	36	40
2·3	5·290	5·336	5·382	5·429	5·476	5·523	5·570	5·617	5·664	5·712	5	9	14	19	23	28	33	38	42
2·4	5·760	5·808	5·856	5·905	5·954	6·003	6·052	6·101	6·150	6·200	5	10	15	20	24	29	34	39	44
2·5	6·250	6·300	6·350	6·401	6·452	6·503	6·554	6·605	6·656	6·708	5	10	15	20	25	31	36	41	46
2·6	6·760	6·812	6·864	6·917	6·970	7·023	7·076	7·129	7·182	7·236	5	11	16	21	26	32	37	42	48
2·7	7·290	7·344	7·398	7·453	7·508	7·563	7·618	7·673	7·728	7·784	5	11	16	22	27	33	38	44	49
2·8	7·840	7·896	7·952	8·009	8·066	8·123	8·180	8·237	8·294	8·352	6	11	17	23	28	34	40	46	51
2·9	8·410	8·468	8·526	8·585	8·644	8·703	8·762	8·821	8·880	8·940	6	12	18	24	29	35	41	47	53
3·0	9·000	9·060	9·120	9·181	9·242	9·303	9·364	9·425	9·486	9·548	6	12	18	24	30	37	43	49	55
3·1	9·610	9·672	9·734	9·797	9·860	9·923	9·986				6	13	19	25	31	38	44	50	57
3·1								10·05	10·11	10·18	1	1	2	3	3	4	4	5	6
3·2	10·24	10·30	10·37	10·43	10·50	10·56	10·63	10·69	10·76	10·82	1	1	2	3	3	4	5	5	6
3·3	10·89	10·96	11·02	11·09	11·16	11·22	11·29	11·36	11·42	11·49	1	1	2	3	3	4	5	5	6
3·4	11·56	11·63	11·70	11·76	11·83	11·90	11·97	12·04	12·11	12·18	1	1	2	3	3	4	5	6	6
3·5	12·25	12·32	12·39	12·46	12·53	12·60	12·67	12·74	12·82	12·89	1	1	2	3	4	4	5	6	6
3·6	12·96	13·03	13·10	13·18	13·25	13·32	13·40	13·47	13·54	13·62	1	1	2	3	4	4	5	6	7
3·7	13·69	13·76	13·84	13·91	13·99	14·06	14·14	14·21	14·29	14·36	1	2	2	3	4	5	5	6	7
3·8	14·44	14·52	14·59	14·67	14·75	14·82	14·90	14·98	15·05	15·13	1	2	2	3	4	5	5	6	7
3·9	15·21	15·29	15·37	15·44	15·52	15·60	15·68	15·76	15·84	15·92	1	2	2	3	4	5	6	6	7
4·0	16·00	16·08	16·16	16·24	16·32	16·40	16·48	16·56	16·65	16·73	1	2	2	3	4	5	6	6	7
4·1	16·81	16·89	16·97	17·06	17·14	17·22	17·31	17·39	17·47	17·56	1	2	2	3	4	5	6	7	7
4·2	17·64	17·72	17·81	17·89	17·98	18·06	18·15	18·23	18·32	18·40	1	2	3	3	4	5	6	7	8
4·3	18·49	18·58	18·66	18·75	18·84	18·92	19·01	19·10	19·18	19·27	1	2	3	4	5	5	6	7	8
4·4	19·36	19·45	19·54	19·62	19·71	19·80	19·89	19·98	20·07	20·16	1	2	3	4	5	5	6	7	8
4·5	20·25	20·34	20·43	20·52	20·61	20·70	20·79	20·88	20·98	21·07	1	2	3	4	5	5	6	7	8
4·6	21·16	21·25	21·34	21·44	21·53	21·62	21·72	21·81	21·90	22·00	1	2	3	4	5	6	7	7	8
4·7	22·09	22·18	22·28	22·37	22·47	22·56	22·66	22·75	22·85	22·94	1	2	3	4	5	6	7	8	9
4·8	23·04	23·14	23·23	23·33	23·43	23·52	23·62	23·72	23·81	23·91	1	2	3	4	5	6	7	8	9
4·9	24·01	24·11	24·21	24·30	24·40	24·50	24·60	24·70	24·80	24·90	1	2	3	4	5	6	7	8	9
5·0	25·00	25·10	25·20	25·30	25·40	25·50	25·60	25·70	25·81	25·91	1	2	3	4	5	6	7	8	9
5·1	26·01	26·11	26·21	26·32	26·42	26·52	26·63	26·73	26·83	26·94	1	2	3	4	5	6	7	8	9
5·2	27·04	27·14	27·25	27·35	27·46	27·56	27·67	27·77	27·88	27·98	1	2	3	4	5	6	7	8	9
5·3	28·09	28·20	28·30	28·41	28·52	28·62	28·73	28·84	28·94	29·05	1	2	3	4	5	6	7	9	10
5·4	29·16	29·27	29·38	29·48	29·59	29·70	29·81	29·92	30·03	30·14	1	2	3	4	6	7	8	9	10

TABLE 9: (cont.) *Squares of Numbers*

	0	1	2	3	4	5	6	7	8	9	1	2	3	4	5	6	7	8	9
5·5	30·25	30·36	30·47	30·58	30·69	30·80	30·91	31·02	31·14	31·25	1	2	3	4	6	7	8	9	10
5·6	31·36	31·47	31·58	31·70	31·81	31·92	32·04	32·15	32·26	32·38	1	2	3	5	6	7	8	9	10
5·7	32·49	32·60	32·72	32·83	32·95	33·06	33·18	33·29	33·41	33·52	1	2	3	5	6	7	8	9	10
5·8	33·64	33·76	33·87	33·99	34·11	34·22	34·34	34·46	34·57	34·69	1	2	4	5	6	7	8	9	11
5·9	34·81	34·93	35·05	35·16	35·28	35·40	35·52	35·64	35·76	35·88	1	2	4	5	6	7	8	10	11
6·0	36·00	36·12	36·24	36·36	36·48	36·60	36·72	36·84	36·97	37·09	1	2	4	5	6	7	9	10	11
6·1	37·21	37·33	37·45	37·58	37·70	37·82	37·95	38·07	38·19	38·32	1	2	4	5	6	7	9	10	11
6·2	38·44	38·56	38·69	38·81	38·94	39·06	39·19	39·31	39·44	39·56	1	3	4	5	6	8	9	10	11
6·3	39·69	39·82	39·94	40·07	40·20	40·32	40·45	40·58	40·70	40·83	1	3	4	5	6	8	9	10	11
6·4	40·96	41·09	41·22	41·34	41·47	41·60	41·73	41·86	41·99	42·12	1	3	4	5	6	8	9	10	12
6·5	42·25	42·38	42·51	42·64	42·77	42·90	43·03	43·16	43·30	43·43	1	3	4	5	7	8	9	10	12
6·6	43·56	43·69	43·82	43·96	44·09	44·22	44·36	44·49	44·62	44·76	1	3	4	5	7	8	9	11	12
6·7	44·89	45·02	45·16	45·29	45·43	45·56	45·70	45·83	45·97	46·10	1	3	4	5	7	8	9	11	12
6·8	46·24	46·38	46·51	46·65	46·79	46·92	47·06	47·20	47·33	47·47	1	3	4	5	7	8	10	11	12
6·9	47·61	47·75	47·89	48·02	48·16	48·30	48·44	48·58	48·72	48·86	1	3	4	6	7	8	10	11	13
7·0	49·00	49·14	49·28	49·42	49·56	49·70	49·84	49·98	50·13	50·27	1	3	4	6	7	8	10	11	13
7·1	50·41	50·55	50·69	50·84	50·98	51·12	51·27	51·41	51·55	51·70	1	3	4	6	7	9	10	11	13
7·2	51·84	51·98	52·13	52·27	52·42	52·56	52·71	52·85	53·00	53·14	1	3	4	6	7	9	10	12	13
7·3	53·29	53·44	53·58	53·73	53·88	54·02	54·17	54·32	54·46	54·61	1	3	4	6	7	9	10	12	13
7·4	54·76	54·91	55·06	55·20	55·35	55·50	55·65	55·80	55·95	56·10	1	3	4	6	7	9	10	12	13
7·5	56·25	56·40	56·55	56·70	56·85	57·00	57·15	57·30	57·46	57·61	2	3	5	6	8	9	11	12	14
7·6	57·76	57·91	58·06	58·22	58·37	58·52	58·68	58·83	58·98	59·14	2	3	5	6	8	9	11	12	14
7·7	59·29	59·44	59·60	59·75	59·91	60·06	60·22	60·37	60·53	60·68	2	3	5	6	8	9	11	12	14
7·8	60·84	61·00	61·15	61·31	61·47	61·62	61·78	61·94	62·09	62·25	2	3	5	6	8	9	11	13	14
7·9	62·41	62·57	62·73	62·88	63·04	63·20	63·36	63·52	63·68	63·84	2	3	5	6	8	10	11	13	14
8·0	64·00	64·16	64·32	64·48	64·64	64·80	64·96	65·12	65·29	65·45	2	3	5	6	8	10	11	13	14
8·1	65·61	65·77	65·93	66·10	66·26	66·42	66·59	66·75	66·91	67·08	2	3	5	7	8	10	11	13	15
8·2	67·24	67·40	67·57	67·73	67·90	68·06	68·23	68·39	68·56	68·72	2	3	5	7	8	10	12	13	15
8·3	68·89	69·06	69·22	69·39	69·56	69·72	69·89	70·06	70·22	70·39	2	3	5	7	8	10	12	13	15
8·4	70·56	70·73	70·90	71·06	71·23	71·40	71·57	71·74	71·91	72·08	2	3	5	7	8	10	12	14	15
8·5	72·25	72·42	72·59	72·76	72·93	73·10	73·27	73·44	73·62	73·79	2	3	5	7	9	10	12	14	15
8·6	73·96	74·13	74·30	74·48	74·65	74·82	75·00	75·17	75·34	75·52	2	3	5	7	9	10	12	14	16
8·7	75·69	75·86	76·04	76·21	76·39	76·56	76·74	76·91	77·09	77·26	2	4	5	7	9	11	12	14	16
8·8	77·44	77·62	77·79	77·97	78·15	78·32	78·50	78·68	78·85	79·03	2	4	5	7	9	11	12	14	16
8·9	79·21	79·39	79·57	79·74	79·92	80·10	80·28	80·46	80·64	80·82	2	4	5	7	9	11	13	14	16
9·0	81·00	81·18	81·36	81·54	81·72	81·90	82·08	82·26	82·45	82·63	2	4	5	7	9	11	13	14	16
9·1	82·81	82·99	83·17	83·36	83·54	83·72	83·91	84·09	84·27	84·46	2	4	5	7	9	11	13	15	16
9·2	84·64	84·82	85·01	85·19	85·38	85·56	85·75	85·93	86·12	86·30	2	4	6	7	9	11	13	15	17
9·3	86·49	86·68	86·86	87·05	87·24	87·42	87·61	87·80	87·98	88·17	2	4	6	7	9	11	13	15	17
9·4	88·36	88·55	88·74	88·92	89·11	89·30	89·49	89·68	89·87	90·06	2	4	6	8	9	11	13	15	17
9·5	90·25	90·44	90·63	90·82	91·01	91·20	91·39	91·58	91·78	91·97	2	4	6	8	10	11	13	15	17
9·6	92·16	92·35	92·54	92·74	92·93	93·12	93·32	93·51	93·70	93·90	2	4	6	8	10	12	14	15	17
9·7	94·09	94·28	94·48	94·67	94·87	95·06	95·26	95·45	95·65	95·84	2	4	6	8	10	12	14	16	18
9·8	96·04	96·24	96·43	96·63	96·83	97·02	97·22	97·42	97·61	97·81	2	4	6	8	10	12	14	16	18
9·9	98·01	98·21	98·41	98·60	98·80	99·00	99·20	99·40	99·60	99·80	2	4	6	8	10	12	14	16	18

Mean differences: columns 1–9

TABLE 10: *Converting a z-Score to Percentile Rank*

z-Score	Percentile rank	z-Score	Percentile rank	z-Score	Percentile rank
		− 1·5	7	0·8	79
		− 1·4	8	0·9	82
		− 1·3	10	1·0	84
		− 1·2	12	1·1	86
		− 1·1	14	1·2	88
		− 1·0	16	1·3	90
		− 0·9	18	1·4	92
		− 0·8	21	1·5	93
− 3·0	0·1	− 0·7	24	1·6	94·5
− 2·9	0·2	− 0·6	28	1·7	95·5
− 2·8	0·3	− 0·5	31	1·8	96·4
− 2·7	0·4	− 0·4	34	1·9	97·1
− 2·6	0·5	− 0·3	38	2·0	97·7
− 2·5	0·6	− 0·2	42	2·1	98·2
− 2·4	0·8	− 0·1	46	2·2	98·6
− 2·3	1·1	0	50	2·3	98·9
− 2·2	1·4	+ 0·1	54	2·4	99·2
− 2·1	1·8	0·2	58	2·5	99·4
− 2·0	2·3	0·3	62	2·6	99·5
− 1·9	2·9	0·4	66	2·7	99·6
− 1·8	3·6	0·5	69	2·8	99·7
− 1·7	4·5	0·6	72	2·9	99·8
− 1·6	5·5	0·7	76	3·0	99·9

APPENDIX 3
Answers

Answers to Chapter Questions
Question
1. rank order $=30$ 29 28 27 25 23 21 20 19 18
   median $=24$
2. mode $=9$
   mean $=5\frac{11}{13}$
3. rank order $=31$ 27 22 19 19 14 11 9 8
   median $=19$
4. statistics, parameters
5. skewed
6. (i) heaven alone knows! (ii) median (iii) modal (iv) below
   the mean (v) modal
7. (b) mean $=207\cdot6$ (c) median $=201\cdot5$
   (d) mode $=189$ 195 200 208. These all appear twice but
   no clear mode or modes have in fact developed.
   (e) and (f) better picture of class average given by the median
   score as the distribution is skewed
8. standard deviation $=0\cdot63$
9. standard deviation $=7\cdot07$
10. standard deviation $=7\cdot906$
11. (a) $+1$ (b) $+2$ (c) $-0\cdot5$ (d) $-1\cdot9$ (e) $+2\cdot2$
12. (a) $+1$ (b) $+2$ (c) -1 (d) -2 (e) $-2\cdot8$ (f) 0
13. (a) $99\cdot9$ (b) $97\cdot7\%$ (c) 50% (d) 16%
14. mean geography $=63\cdot95$
    sigma geography $=16\cdot0$
    mean history $=68\cdot5$
    sigma history $=3\cdot38$
    Percy did better at history because he was further above the
    mean for that subject in terms of standard deviations.
15. Depending on the size of the intervals, answers will vary slightly.
    The following ranges are acceptable:
    (c) second decile between 35 and 37, third quartile between
    64 and 66, 95th percentile between 82 and 83

    (d) pass mark $= 47\%$, credit $= 73\%$

16. (a) $SE_{mean} = 1\cdot53$
    (b) $SE_{meas} = 1\cdot08$
    (c) $SE_{est} = 1\cdot51$
    (d) Charlie's mark would fall between 18 and 24 for the 95% confidence level (i.e. $\pm 2\ SE_{est}$)

17. Spearman's $r = -0\cdot30$

18. $t = 0\cdot35$, therefore difference between boys and girls is not significant

19. (a) 37·3   (b) 50   (c) 53·4

20. Highest multiple R (0·4123) is between heart attack, height and family history

21. Effect of social background on GCE results when IQ is held constant $= + 0\cdot42$

22. There is no significant difference between the sub-groups and the whole population.

23. The difference is not significant

24. Your result should be close to a mean of 234 and sigma 25

25. Pearson's r is approximately 0·46 depending on the size of interval

Answers to Further Problems

Question

1. mean IQ $= 108\cdot7$
2. median IQ $= 104$
3. day no. 183 $= 2$ July
4. mode $= 31$
5. sigma $= 17\cdot00$
6. sigma $= 17\cdot24$
7. z-scores would be 44·88, 41·94, 36·65, 51·35, 52·53, 27·82
8. The scores are fairly evenly distributed, with slight evidence of left-handed skew
9. (a) SE_{mean} classics $= 3\cdot543$, modern languages 5·247, English literature 4·268, arts 3·356
    (b) Scores vary over wide range. The size of samples is too small to expect much useful prediction from scores.

10. (a) $SE_{meas} = 4.98$

  (b) Range likely for 0.05 level of confidence is 2 SE_{meas}, i.e. 109.04 to 128.96

  (c) $SE_{est} = 6.84$

  (d) Range likely for 0.05 level of confidence is 2 SE_{est}, i.e. 105.32 to 132.68

11. (a) No! Spearman's formula is generally held to be inaccurate with groups larger than 30

  (b) They are all significant beyond the 0.01 level

12. Spearman's $r = +0.04$

13. Pearson's $r = 0.7$

14. Partial tetrachoric $r = -0.31$

15. The difference is not significant

16. The difference is significant beyond the 0.01 level

17. McCall T-scores are as follows:

Student	Education mark	Teaching mark	Bio. mark	Eng. mark	Hist. mark	Geog. mark	Total
1	56.96	57.45	36	—	—	—	150.41
2	61.42	57.45	54	—	—	—	172.87
3	65.88	56.27	60	—	—	—	182.15
4	35.79	56.27	—	41	—	—	133.06
5	38.02	45.08	—	46	—	—	129.10
6	52.51	68.64	—	64	—	—	185.15
7	56.96	57.45	—	—	61	—	175.41
8	59.19	56.27	—	—	55	—	170.46
9	48.05	56.27	—	—	37	—	141.32
10	44.71	45.08	—	—	—	62	153.79
11	34.68	56.27	—	—	—	49	139.95
12	45.82	56.27	—	—	—	39	141.09

Biology, English, History and Geography marks have been rounded off. Student No. 6 should get the over-all academic prize.

18. $R = 0.095$

19. None of the subgroups differ significantly from the population, or from each other

20. Significance is beyond the 0.01 level

APPENDIX 4
An objective test

YOU SHOULD HAVE a question sheet and an answer sheet. Please do NOT write on the question sheet. If you need to scribble do so on the back of your answer sheet. Leave your question sheet and answer sheet behind when you go. You have plenty of time, do not rush. Make sure that you have chosen an answer for *every* question because there is *no* guessing correction being made. Please place your tick *IN* the box of your choice. Only one tick should appear for each question. Read carefully.

TRUE/FALSE

1. Extreme scores, such as a very high or a very low score, distort the median.
2. In a perfectly normal distribution the mean, median and mode are at exactly the same point.
3. The mode is the most frequently occurring score.
4. Partial tetrachoric correlations use 'N' equals half of the class size.
5. A test which is not valid is useful provided it is reliable.
6. For a given sample we can learn as much from a negative correlation as from a positive correlation of the same size.
7. A correlation co-efficient of $+ 0.19$ is always insignificant.
8. The standard deviation of tests is always 15.
9. The range is a measure of the spread of a set of scores.
10. The standard deviation is a measure of the spread of a set of scores.

Multiple Choice

11. If the mean is 25 and the standard deviation of 0.5 what percentage of scores will be between 23.50 and 26.50?
    (a) 34% (b) 68% (c) 95% (d) 99.7%
12. Given a set of scores which fit the standard normal distribution, having a mean of 20 and a standard deviation of 5, between which limits will 68% of the scores lie?

(a) 15–30   (b) 20–30   (c) 15–25   (d) 10–30

13. On learning that a pupil has an IQ of 145 you should

   (a) know his score was three standard deviations above the mean
   (b) know he was a gifted pupil
   (c) know both (a) and (b) above
   (d) Ask what are the test mean and standard deviation.

14. Standardizing our scores enables us to

   (a) balance marks from various examiners
   (b) compare a pupil's marks with an original sample of marks
   (c) both (a) and (b) above
   (d) neither (a) nor (b) above

15. Test 'X' has a mean of 80 and a standard deviation of 10
Test 'Y' has a mean of 90 and a standard deviation of 5
Freddy scored 100 on test 'X'. Johnny scored 100 on test 'Y'.
Who did the better

   (a) Johnny
   (b) Freddy
   (c) They did as well as each other
   (d) Cannot tell from data supplied.

16. Given a choice of Formula

$$SD = \sqrt{\frac{\sum D^2}{N-1}} \text{ or } SD = \sqrt{\frac{\sum D^2}{N}}$$

choose the more appropriate and calculate the standard deviation
of the following scores ($N = 11$)
6, 7, 7, 5, 9, 8, 7, 7, 7, 7, 7,
(NB: Really neither Formula fits as the scores are so far from
being normally distributed.)
(a) − 1·00   (b) 0·00   (c) 0·91   (d) 1·00

17. Reliability correlations

   (a) are a measure of a test's consistency
   (b) are in effect the same as validity correlations
   (c) are always lower than validity correlations
   (d) are always higher than validity correlations.

BASIC STATISTICS OBJECTIVE TEST

ANSWER SHEET

SURNAME (Please print) ....

CHRISTIAN NAMES..

True/False   *Multiple Choice*

1	T	
	F	
2	T	
	F	
3	T	
	F	
4	T	
	F	
5	T	
	F	
6	T	
	F	
7	T	
	F	
8	T	
	F	
9	T	
	F	
10	T	
	F	

11	A	
	B	
	C	
	D	
12	A	
	B	
	C	
	D	
13	A	
	B	
	C	
	D	
14	A	
	B	
	C	
	D	
15	A	
	B	
	C	
	D	

16	A	
	B	
	C	
	D	
17	A	
	B	
	C	
	D	

Score = ...

BASIC STATISTICS OBJECTIVE TEST
ANSWER KEY

SURNAME (Please print) ....

CHRISTIAN NAMES...

True/False

1	T	
	F	▨
2	T	▨
	F	
3	T	▨
	F	
4	T	
	F	▨
5	T	
	F	▨
6	T	▨
	F	
7	T	
	F	▨
8	T	
	F	▨
9	T	▨
	F	
10	T	▨
	F	

Multiple Choice

11	A	
	B	
	C	
	D	▨
12	A	
	B	
	C	▨
	D	
13	A	
	B	
	C	
	D	▨
14	A	▨
	B	
	C	
	D	
15	A	
	B	
	C	▨
	D	

16	A	
	B	
	C	
	D	▨
17	A	▨
	B	
	C	
	D	

Score = ▨

*Cut out shaded squares
so that correct responses
show through*

REFERENCES

AHMANN, J. S. and GLOCK, M. D. (1963), *Evaluating Pupil Growth*, Allyn & Bacon.

AMOS, J. R., BROWN, G. and MINK, O. G. (1965), *Statistical Concepts*, Harper & Row.

BARROW, H. M. and McGEE R. (1954), *Measurement in Physical Education*, Lea & Febiger.

CRONBACH, L. J. (1961), *Essentials of Psychological Testing*, Harper & Row.

DUPOIS, P. H. (1965), *An Introduction to Psychological Statistics*, Harper & Row.

EYSENCK, H. J. (1971), *Race, Intelligence and Education*, Temple Smith.

FARROW, F. (1962), *Statistical Measures*, Chandler.

FISHER, R. A. (1963), *Statistical Methods for Research Workers*, Hafner, 13th edn.

FISHER, R. A. and YATES, F. (1963), *Statistical Tables for Biological, Agricultural and Medical Research*, Oliver & Boyd, 6th edn.

KELLAWAY, F. W. (ed.) (1968), *The Penguin–Honeywell Book of Tables*, Penguin.

LINDQUIST, E. F. (1940), *Statistical Analysis in Educational Research*, Houghton Mifflin.

LINDQUIST, E. F. (1951), *Educational Measurement*, American Council on Education.

LYMAN, A. B. (1963), *Test Scores and What They Mean*, Prentice-Hall.

REICHMANN, W. J. (1964), *Use and Abuse of Statistics*, Penguin.

SPIEGAL, M. R. (1961), *Statistics*, Schaum.

WUNDT, E. and BROWN, G. (1957), *Essentials of Educational Measurement*, Holt, Rinehart & Winston.

YEOMANS, K. A. (1968), *Statistics for the Social Scientist*, vol. 1: *Introducing Statistics*, Penguin.

YEOMANS, K. A. (1968), *Statistics for the Social Scientist*, vol. 2: *Applied Statistics*, Penguin.

Reference has also been made to articles appearing in the following journals:

Educational Research—November 1964, November 1965, February 1966, November 1966, June 1967.

British Journal of Educational Psychology—February 1948.

INDEX OF FORMULAE (In order of first appearance)

Formula	Page	Pre-requisite statistical knowledge	Used to calculate
$M = \dfrac{\sum x}{N}$	19	Nil	The mean (arithmetic average) of a set of scores.
$\sigma = \sqrt{\left(\dfrac{\sum d^2}{N}\right)}$	23	How to calculate the mean	Standard Deviation. A measure of how scores are spread out. Used for populations. Basic part of many more complex formulae
$SD = \sqrt{\left(\dfrac{\sum d^2}{N-1}\right)}$	25	How to calculate the mean	Standard deviation of small samples of less than 30 scores. Also to calculate standard deviation of population when only a sample is used
$R = \dfrac{2r}{1+r}$	43	Either know the split half correlation figure or be able to calculate same	The full test reliability figure when test is able to be split into two equal parts
$SE_{meas} = \sigma \sqrt{(1-r)}$	49	How to calculate standard deviations and correlation coefficients	The standard error of measurement, i.e. the amount of error likely to be involved in a person's score on a particular test

Page	Topic	Formula	Description
50	How to calculate a standard deviation	$SE_{mean} = \dfrac{SD}{\sqrt{N}}$	The standard error of the mean, i.e. how much the sample mean can be expected to vary from the population mean
51	How to calculate a standard deviation and correlation coefficient	$SE_{est} = \sigma\sqrt{(1 - r^2)}$	The standard error of estimate, i.e the amount of error likely to be involved when predicting a future score on a test from a current score on the same test
58	How to calculate the mean	$r = 1 - \dfrac{6\sum d^2}{N(N^2 - 1)}$	Spearman's rank order correlation is used for comparing scores of small samples or populations of less than 30 to see whether there is any relationship
60	Nil	$\dfrac{q}{N} \times 100\%$	The partial tetrachoric correlation between two sets of scores. A rough quick method suitable for more than 30 sets of scores
61	Nil	$r = \dfrac{\sum xy - \dfrac{(\sum x)(\sum y)}{N}}{\sqrt{\left[\sum x^2 - \dfrac{(\sum x)^2}{N}\right]\left[\sum y^2 - \dfrac{(\sum y)^2}{N}\right]}}$	Pearson's Product Moment correlations. A well-respected, much used method suitable for large or small sets of scores when wanting to find the relationship between the two

Formula	Page	Prerequisite	Description
$t = \dfrac{M_1 + M_2}{\sqrt{\left[\left(\dfrac{\sum d_1^2 + \sum d_2^2}{N_1 + N_2 - 2}\right)\left(\dfrac{1}{N_1} = \dfrac{1}{N_2}\right)\right]}}$	68	Able to compute mean	t-Test of significance which is used when comparing two small groups of data to see if they are significantly different
$CR = \dfrac{M_2 - M_1}{\sqrt{[(SE_{mean})^2 + (SE_{mean})^2]}}$	71	Ability to calculate mean standard deviation and standard error mean	Significance of difference of means Used when comparing larger groups of data to see if they are significantly different
$Discrim. = \dfrac{H - L}{N}$	76	Nil	Item discrimination is used to help decide whether an objective question is good or poor, e.g. a good one will be got right by those testees doing better on whole test
$Diffic. = \dfrac{Right}{N} \times 100\%$	77	Nil	Item difficulty. How hard an objective question is
$z\text{-Score} = \dfrac{d}{\sigma}$	81	Ability to calculate mean and standard deviation	Standardized z-scores
$T\text{-score} = 50 \pm 10\left(\dfrac{d}{\sigma}\right)$	81	Ability to calculate mean and standard deviation	McCall T-scores based on a mean of 50 and σ of 10
$R_{1.23} = \sqrt{\left[\dfrac{r_{12}^2 + r_{13}^2 - 2(r_{12} \times r_{13} \times r_{23})}{1 - r_{23}^2}\right]}$	83	Ability to calculate zero order correlation coefficients	Multiple correlations for predicting criterion from two zero order predictors

Formula	Requirement	Page	Description
$r_{12\cdot3}=\dfrac{r_{12}-r_{13}\,r_{23}}{\sqrt{(1-r_{13}^2)(1-r_{23}^2)}}$	Ability to calculate zero order correlation coefficients	84	Enables the relationship between two variables to be calculated with the effect of a third held constant
$F\text{-ratio}=\dfrac{\text{between variance}}{\text{within variance}}$	Able to calculate between variance and within variance (see next item)	87	Variance allowing comparison of several groups to see whether a significant difference exists
Between variance $=\dfrac{\sum(d^2{}_{\text{mean}})}{df_{(\text{groups}-1)}}$	Ability to calculate mean	87	Needed for above
Within variance $=\dfrac{\sum d^2}{df_{(N-\text{groups})}}$		87	
$x^2=\sum\sum\left[\dfrac{(A-E)^2}{E}\right]$	Nil	90	Chi square which allows us to test an hypothesis and so decide whether it is significantly valid or not
Act. mean $=$ Ass. mean $+\;i\,\dfrac{\sum fd}{N}$	Nil	96	Mean via grouped data
$\sigma=\left\{\sqrt{\left[\dfrac{\sum fd^2}{N}-\left(\dfrac{\sum fd}{N}\right)^2\right]}\right\}i$	Nil	97	Standard deviation via grouped data
$r=\dfrac{\dfrac{\sum(f_1 d_1\checkmark)}{N}-\left(\dfrac{\sum f_1 d_1}{N}\right)\left(\dfrac{\sum f_2 d_2}{N}\right)}{\sigma f_1\times\sigma f_2}$	Nil	103	Pearson product moment correlation via grouped data

INDEX OF POSSIBLE NEEDS

Requirement	Provided by	Chapter
Selection of a sample unaffected by sampling bias	Reading chapter on sampling, etc.	14
Average of a set of scores	Calculating mean, mode or median	1 13
Measure of spread of a set of scores	Calculating range, standard deviation	2 13
Means of offsetting examiner's marking bias	Calculating z-scores, McCall T-scores	2 9
Should a test question be kept or discarded?	Calculating item discrimination (item analysis)	8
Want test questions in ascending order of difficulty	Calculating item difficulty (item analysis)	8
Is a test reliable?	Calculating split half, parallel form or test-retest correlation coefficient	4
Is a predictive test valid?	Calculating zero order correlations or chi square	4
How much can a person's test score be trusted to be accurate?	Calculating standard error of measurement	5
How much trust to place in a score's ability to predict future score on same test?	Calculating standard error of estimate	5

Requirement	Provided by	Chapter
How representative is a sample of the population from which it's drawn?	Calculating standard error of the mean	5
Amount of relationship between two sets of scores for one group of people	Calculating rank order, partial tetrachoric or product moment correlation coefficient	6 13
To use two sets of information to predict a third	Calculating 1st order multiple coefficient of correlation	10
Is a correlation coefficient large enough to be significant	Look up Table 3a or Table 3b	Appendix 2
Comparison of expected results with actual to see whether significantly different	Calculating chi square	12
Comparison of two small groups to see whether they differ significantly	Calculating t-test of significant difference or Tukey's test	7
Comparison of two large groups to see whether they differ significantly	Calculating significance of difference between the means	7
Comparison of several groups to see if any one differs significantly	Calculating the F-ratio	11
Finding relationship between two factors' whilst holding the effect of a third constant	Partial correlation	10
Constructing an objective test	Test construction	8

GENERAL INDEX

(This replaces the existing index, which is incorrect)